T0231391

IMPEDANCE BOUNDARY CONDITIONS IN ELECTROMAGNETICS

ELECTROMAGNETICS LIBRARY
C. E. BAUM, Editor

Baum and Stone
Transient Lens Synthesis: Differential Geometry in Electromagnetic Theory

Bowman, Senior, and Uslenghi
Electromagnetic and Acoustic Scattering by Simple Shapes

Gardner
Lightning Electromagnetics

Hoppe and Rahmat-Samii
Impedance Boundary Conditions in Electromagnetics

Lee
EMP Interaction: Principles, Techniques, and Reference Data

Mittra
Computer Techniques for Electromagnetics

Smythe
Static and Dynamic Electricity

Taylor and Giri
High-Power Microwave Systems and Effects

Van Bladel
Electromagnetic Fields

IN PREPARATION

Baum and Kritikos
Electromagnetic Symmetry

IMPEDANCE BOUNDARY CONDITIONS IN ELECTROMAGNETICS

Daniel Jay Hoppe, Ph.D.

Yahya Rahmat-Samii, Ph.D.

University of California at Los Angeles

CRC Press
Taylor & Francis Group
Boca Raton   London   New York

CRC Press is an imprint of the
Taylor & Francis Group, an informa business

USA	Publishing Office:	Taylor & Francis 1101 Vermont Avenue, N.W., Suite 200 Washington, DC 20005-2531 Tel: (202) 289-2174 Tax: (202) 289-3665
	Distribution Center:	Taylor & Francis 1900 Frost Road, Suite 101 Bristol, PA 19007-1598 Tel: (215) 785-5800 Fax: (215) 785-5515
UK		Taylor & Francis Ltd 4 John Street London WC1N 2ET, UK Tel: 071 405 2237 Fax: 071 831 2035

IMPEDANCE BOUNDARY CONDITIONS IN ELECTROMAGNETICS

Cover design by Michelle Fleitz.
A CIP catalog record for this book is available from the British Library.
⊚ The paper in this publication meets the requirements of the ANSI Standard Z39.48-1984(Permanence of Paper)

Library of Congress Cataloging-in-Publication Data

Hoppe, Daniel Jay.
    Impedance boundary conditions in electromagnetics / by Daniel Jay Hoppe, Yahya Rahmat-Samii.
    p. cm.
    Includes bibliographical references and index.
    1. Electromagnetic waves—Scattering—Methodology. 2. Boundary value problems. I. Rahmat-Samii, Yahya. II. Title.
QC665.S3H68   1995
621.3848—dc20   94-42315
ISBN 1-56032-385-X   CIP

Reprinted 2010 by CRC Press
CRC Press
6000 Broken Sound Parkway, NW
Suite 300, Boca Raton, FL 33487
270 Madison Avenue
New York, NY 10016
2 Park Square, Milton Park
Abingdon, Oxon OX14 4RN, UK

Contents

Preface

Electromagnetic scattering from complex objects has been an area of in-depth research for many years. A variety of powerful solution methodologies have been developed and utilized for the clever solution of ever-increasingly complex problems. Among these methodologies, the subject of impedance boundary conditions has interested the authors for some time. In short, impedance boundary conditions allow one to replace a complex structure with an appropriate impedance relationship between the electric and magnetic fields on the surface of the object. This simplifies the solution of the problem considerably, allowing one to ignore the complexity of the internal structure beneath the surface. Through this book we hope to demonstrate the inherent beauty and elegance of impedance boundary conditions in electromagnetics.

The reader is referred to Chapter 1 "Introduction" for a detailed presentation of the role of the impedance boundary conditions in solving practical electromagnetic problems, some historical background, and the organization of this book. One of the main objectives of this book is to present a unified and thorough discussion of this important subject. To our best knowledge there is not a single book available that is entirely devoted to the subject of impedance boundary conditions in electromagnetics. In this book a novel and unified method based on a spectral domain approach is presented to derive the Higher Order Impedance Boundary Conditions (HOIBC). The method includes all of the existing approximate boundary conditions, such as the Standard Impedance Boundary Condition, the Tensor Impedance Boundary Condition, and the Generalized Impedance Boundary Conditions, as special cases.

The spectral domain approach is applicable to complex coatings and surface treatments as well as simple dielectric coatings. The spectral domain approach is employed to determine the appropriate boundary conditions for planar dielectric coatings, chiral coatings, and corrugated conductors. The accuracy of the proposed boundary conditions is discussed. The approach is then extended to include the effects of curvature and is applied to curved dielectric and chiral coatings.

After a presentation of the general theory, the application of Higher Order Impedance Boundary Conditions (HOIBC) to scattering by several representative configurations is considered. In each case the HOIBC solutions are compared to closed form solutions or numerical solutions of integral equations that are based on an exact formulation. In all cases the HOIBC solutions are found to reduce computation time and resource requirements relative to the exact formulation.

Among the geometries considered are dielectric-filled grooves, coated conducting cylinders of arbitrary shape, and coated bodies of revolution. Ample numerical data are presented to critically assess the accuracy of the results obtained using various forms of the impedance boundary conditions. A number of appendices are included in order to provide more detail on some of the topics addressed in the main body of the book. A selective list of references directly related to the topics addressed in this book is included.

It is believed that the material presented in this book should be of use to researchers in the areas of electromagnetic scattering and advanced solution techniques in electromagnetics as well as students pursuing their studies in the area of advanced electromagnetics. The book can also be used as a supplemental book for graduate courses in electromagnetics.

Daniel Jay Hoppe
Yahya Rahmat-Samii

Nomenclature

English Symbols

a	inner radius of coated circle	[m]
a	axis dimension of superquadric cylinder	[m]
$\{a\}$	vector of forward wave amplitudes	[1]
$\{a^\alpha\}$	vector of BOR electric currents	[A]
$a(i)^m$	amplitude of ith forward mode in mth wave-guide	[1]
$\hat{a}_1$	unit tangent along the type 1 BOR curve	[1]
$\hat{a}_n$	unit normal	[1]
$\hat{a}_t$	unit tangent	[1]
$a_n^t(j)$	amplitude of t directed electric current on jth triangle for nth Fourier component	[A]
$a_n^\phi(j)$	amplitude of ϕ directed electric current on jth triangle for nth Fourier component	[A]
A_{TE}	Magnitude of incident TE plane wave	[V/m]
A_{TM}	Magnitude of incident TM plane wave	[V/m]
b	outer radius of coated circle	[m]
b	axis dimension of superquadric cylinder	[m]
$\{b\}$	vector of backward wave amplitudes	[1]
$\{b^\alpha\}$	vector of BOR magnetic current unknowns	[V]
$b(i)^m$	backward wave amplitude of ith mode in mth waveguide	[1]
$b_n^t(j)$	amplitude of t directed magnetic current on jth triangle for nth Fourier component	[A]
$b_n^\phi(j)$	amplitude of ϕ directed magnetic current on jth triangle for nth Fourier component	[A]
$\mathbf{B}$	magnetic flux vector	[Vs/m^2]
$[B]$	overall boundary condition matrix	[Ω]
$[B_J]$	boundary condition matrix	[Ω]
$[B_M]$	boundary condition matrix	[1]

B_{TE}	Magnitude of reflected TE plane wave	[V/m]
B_{TM}	Magnitude of reflected TM plane wave	[V/m]
c_i	wave function amplitude	[1]
c_m^n	complex polynomial coefficient	[−]
$\{C_{12}\}$	coefficient vector	[1]
d	coating thickness	[m]
d	corrugation depth	[m]
ds	elemental surface area	[m²]
dt	elemental arc length	[m]
$\mathbf{D}$	displacement field vector	[As/m²]
$\mathrm{Det}[[Z]]$	determinant of matrix [Z]	[−]
e	a constant (2.718…)	[1]
$e_i(x)$	electric field for the ith parallel plate mode	[V/m]
$\mathbf{e}_i(k_x, k_y)$	rectangular electric field wave function	[V/m]
$\mathbf{e}_i(n, k_z)$	cylindrical electric field wave function	[V/m]
$\mathbf{E}$	electric field vector	[V/m]
$\mathbf{E}_i$	incident electric field vector	[V/m]
$\mathbf{E}(k_x, k_y, z)$	rectangular transverse electric field vector	[V/m]
$\mathbf{E}(k_z, r)$	cylindrical transverse electric field vector	[V/m]
$E_x^i(x)$	total electric field in ith layer	[V/m]
$\tilde{E}_x(k_x, k_y)$	Fourier transform of E_x	[V/m]
$\{E(z)\}$	transverse electric field vector	[V/m]
$[E_{12}(z)]$	transverse electric field matrix	[V/m]
$F_k(i, j)$	interaction integral between triangles i and j	[1]
$G(\mathbf{r}, \mathbf{r}')$	two dimensional free space Green's function	[1]
$G_0(\mathbf{r}, \mathbf{r}')$	three dimensional free space Green's function	[1/m]
$h_i(x)$	magnetic field for the ith parallel plate mode	[A/m]
$\mathbf{h}_i(k_x, k_y)$	rectangular magnetic field wave function	[A/m]
$\mathbf{h}_i(n, r)$	cylindrical magnetic field wave function	[A/m]
$\mathbf{H}$	magnetic field vector	[A/m]
$\mathbf{H}_i$	incident magnetic field vector	[V/m]
$\mathbf{H}(k_x, k_y, z)$	rectangular transverse magnetic field vector	[A/m]
$\mathbf{H}(k_z, r)$	cylindrical transverse magnetic field vector	[A/m]
$\{H(z)\}$	transverse magnetic field vector	[A/m]
$\tilde{H}_x(k_x, k_y)$	Fourier transform of H_x	[A/m]
$\{H_z\}$	vector of unknown aperture magnetic fields	[A/m]
$H_z^i(x)$	total magnetic field in the ith layer	[A/m]
$H_z^i(x)$	incident magnetic field	[A/m]
$\tilde{H}_z^i(x)$	image of incident magnetic field	[A/m]
$H_z^S(x)$	scattered magnetic field	[A/m]
$H_z^{Tot}(x)$	total magnetic field	[A/m]
$[H_{12}(z)]$	transverse magnetic field matrix	[A/m]
$H_n^{(2)}(x)$	Hankel function of the second kind order n	[1]
$H_n^{(2)'}(x)$	derivative of the Hankel function	[1]

$\{I\}$	source field vector	[V/m]
$[I]$	unit matrix	[1]
$\mathrm{Im}[z]$	imaginary part of z	[−]
j	imaginary number ($\sqrt{-1}$)	[1]
$\mathbf{J}$	electric surface current	[A/m]
$J_t(t)$	tangential electric surface current	[A/m]
J_T	total number of triangles	[1]
$J_n(x)$	Bessel function of the first kind order n	[1]
$J_n'(x)$	derivative of the Bessel function	[1]
k	wavenumber	[1/m]
k_0	free space wavenumber	[1/m]
k_i	wavenumber in the ith layer	[1/m]
k_l	LCP wavenumber	[1/m]
k_r	RCP wavenumber	[1/m]
k_t	azimuthal wave number	[1/m]
(k_x, k_y)	transverse wave numbers	[1/m]
k_y^i	transverse wave number in the ith layer	[1/m]
$k_z^{(i)}$	longitudinal wavenumber	[1/m]
$K[\mathbf{X}(\mathbf{r}')]$	operator in the BOR solution	[−]
l_1	arc length along the type 1 curve	[m]
l_n	total arc length of the nth curve	[m]
$L[\mathbf{X}(\mathbf{r}')]$	operator in the BOR solution	[−]
$\mathbf{M}$	magnetic surface current	[V/m]
$[M]$	matrix used to determine coefficients c_m^n	[−]
$\{M\}$	unknown magnetic current vector	[V/m]
$M_x(t)$	longitudinal magnetic surface current	[V/m]
$M_z(x)$	magnetic surface current	[V/m]
n	Fourier mode number	[1]
$\hat{n}$	unit normal vector	[1]
P	power	[W]
P_0	point on the surface of the BOR	[1]
$[P]$	mode interaction matrix	[1]
$[P_n]$	polynomial matrix in the BOR solution	[−]
$[P^T]$	transpose of mode interaction matrix	[1]
$P_i(x)$	ith pulse function	[1]
$P_n(k_x, k_y)$	polynomial in the transform variables (k_x, k_y)	[−]
$[Q]$	mode interaction matrix	[1]
$\mathrm{Re}[z]$	real part of z	[−]
R_{TETE}	reflection coefficient	[1]
$[S]$	total scattering matrix	[1]
$[S_{ij}]$	submatrix in the total scattering matrix	[1]
S_i	ith surface	[1]
$S_i(t)$	ith spline function	[1]
S_n^-	just inside surface n	[1]

t	eigenvalue determining edge behavior	[1]
t_i	initial tangential coordinate of the ith subdivision	[m]
t_l	eigenvalue for edge behavior at left edge of groove	[1]
t_r	eigenvalue for edge behavior at right edge of groove	[1]
$\hat{t}$	unit tangent vector	[1]
$T_i(t)$	ith triangle function	[1]
$\dot{T}_i(t)$	derivative of the ith triangle function	[1]
$\{V^\alpha\}$	BOR source vector	[V]
w_i	width of ith layer	[m]
$[W]$	partial MOM matrix	[m]
$[W^T]$	overall MOM matrix	[m]
$[W^T]^{-1}$	inverse of overall MOM matrix	[m]
$\mathbf{W}_n^t(i)$	ith tangential weighting function for the nth Fourier component	[1/m]
$\mathbf{W}_n^\phi(i)$	ith azimuthal weighting function for the nth Fourier component	[1/m]
$\hat{x}$	unit vector	[1]
x_l	coordinate of left edge of groove	[m]
x_r	coordinate of right edge of groove	[m]
$\hat{y}$	unit vector	[1]
$[Y]$	MOM matrix	[1]
$[Y]$	admitance matrix	[m/Ω]
$[\tilde{Y}]$	normalized admitance matrix	[1]
$[Y_1]$	partial admitance matrix	[1]
$[Y_2]$	partial admitance matrix	[1]
$[Y^{Tot}]$	total admitance matrix	[1]
z^*	complex conjugate of z	[−]
$z_i(t)$	parametric representation of the z coordinate on the ith BOR surface	[m]
$[Z]$	impedance matrix	[Ω]
$[Z_\gamma^{\alpha\beta}]$	BOR impedance submatrix	[−]
Z_{xx}	term in exact impedance matrix	[Ω]
$\tilde{Z}_{xx}$	term in approximate impedance matrix	[Ω]
$Z_{xx}^{sw}(k_x)$	term in surface wave impedance matrix	[Ω]
Z_{in}^i	input impedance at the ith layer	[Ω]
$Z_{in}|_x(k_x)$	input impedance at groove position x for transverse wavenumber k_x	[Ω]
$\{0\}$	zero vector	[−]

Greek Symbols

α	parameter in the combined field integral equation	[1]
α	surface wave attenuation coefficient	[1/m]
β	propagation constant	[1/m]
γ	superquadric cylinder parameter	[1]
γ	Euler's constant (1.781...)	[1]
γ_c	chiral admittance	[1/Ω]
γ_i	propagation constant for the ith parallel plate mode	[1/m]
γ_p	angle of the pth triangle segment with respect to the z axis	[1]
δ_i	depth of the ith layer	[m]
δ_{pq}	Kroneker delta function	[1]
Δ	pulse width	[m]
ϵ	permittivity	[F/m]
ϵ_0	free space permittivity	[F/m]
ϵ_c	effective permittivity for chiral media	[F/m]
ϵ_i	permittivity of the ith layer	[F/m]
ϵ_r	relative permittivity	[1]
η	wave impedance	[Ω]
η_0	free space wave impedance	[Ω]
η_c	effective wave impedance for chiral media	[Ω]
θ_i	angle of incidence	[1]
μ	permeability	[H/m]
μ_0	free space permeability	[H/m]
μ_i	permeability of the ith layer	[H/m]
μ_r	relative permeability	[1]
π	a constant (3.14159...)	[1]
ρ	distance from groove edge	[m]
$\rho_i(t)$	parametric representation of the radius of the ith BOR surface	[m]
ϕ	azimuthal angle	[1]
$\hat{\phi}$	unit vector	[1]
ϕ_0	angle of incidence	[1]
Φ_i	ith wedge angle	[1]
Ψ	surface wave vector potential	[1]
ω	angular frequency	[1/s]
∞	infinity	[−]
∇	differential (Del) operator	[1/m]

Abbreviations

BOR	Body Of Revolution
CFIE	Combined Field Integral Equation
EFIE	Electric Field Integral Equation
GO	Geometrical Optics
GTD	Geometrical Theory of Diffraction
HOIBC	Higher Order Impedance Boundary Condition
MFIE	Magnetic Field Integral Equation
MOM	Method Of Moments
PO	Physical Optics
PTD	Physical Theory of Diffraction
RCS	Radar Cross Section
SIBC	Standard Impedance Boundary Condition
TIBC	Tensor Impedance Boundary Condition

Chapter 1

Introduction

1.1 Background

The calculation of the scattering of electromagnetic waves by arbitrary bodies
has been an area of research for many decades. Interest in this topic continues
due to the large number of applications of electromagnetic scattering. The
most obvious practical application of electromagnetic scattering is radar [1]. In
this case a body is illuminated by an electromagnetic wave, and the wave is
scattered and received by either the illuminating antenna (monostatic radar)
or by an antenna at another location (bistatic radar);   see Figure 1.1. The
radar cross section (RCS) of the body is a measure of the amount of energy
scattered in a particular direction for a given illumination. In the case of stealth
technology, the objective is to design the body in such a way as to minimize the
amount of energy scattered in a particular direction. Reflections are minimized
by choosing the shape of the body, to the extent possible, and applying surface
coatings or treatments to various portions of the scatterer. Thus, the calculation
of radar cross section is important in the design of potential radar targets, as
well as in the design of the radar system itself. The calculation of radar cross
section for bodies with complex surface treatments or coatings, rather than
simple metal bodies, has become an important area of research in recent years.

Electromagnetic scattering by complicated bodies is often important in the
design of antennas. In many applications support structures or other scattering

1

Figure 1.1: Monostatic and bistatic radar.

bodies are in close proximity to a radiating structure. The presence of the scattering body affects both the radiation pattern of the antenna as well as its impedance characteristics. One example would be the dielectric nose cone of a fighter aircraft affecting the impedance characteristics and radiation pattern of a direction-finding antenna array located directly behind the nose. A second example would be a dual reflector antenna, as shown in Figure 1.2. Here the struts used to support the subreflector block a portion of the radiation reflected off the main reflector, degrading the radiation pattern of the antenna. It has been shown that the degradation of the radiation pattern caused by the struts can be reduced by applying a surface treatment or coating to the conventional metal strut [2].

The above examples illustrate that the calculation of electromagnetic scattering by bodies with complex surface treatments or coatings is an important area of research. While the problem of scattering by perfectly conducting objects is a relatively simple one, the addition of coatings or surface treatments complicates the problem significantly. Even in the era of the supercomputer one must often resort to approximation methods to solve the problem of electromagnetic scattering by these more complicated bodies.

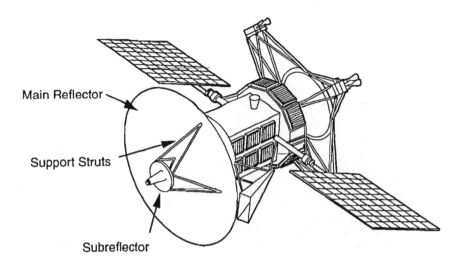

Figure 1.2: A dual reflector antenna.

1.2 Scattering by Conducting Bodies

Consider first the problem of electromagnetic scattering by a perfectly conducting body for some incident electromagnetic field $\mathbf{E}_i$, $\mathbf{H}_i$ in Figure 1.3. The incident field will produce some electric current $\mathbf{J}$ on the body, which will in turn radiate the scattered field. The total tangential electric field, incident plus scattered field, must be zero on the perfectly conducting body. Using the free space Green's function, which relates the scattered electric field to the electric current, an integral equation may be obtained for the electric current on the body. If computer resources are sufficient a solution for the electric current may be obtained, using a numerical procedure such as the method of moments [3]. In general, at least 10 unknowns are required per wavelength for each component of the surface current to obtain accurate results. Thus, approximately 200 unknowns are required for every square wavelength of surface area on the conducting body. Therefore, even for the conducting scatterer a numerical solution based on an exact formulation is possible only for objects of small to moderate size.

For larger conducting objects approximation methods based on optical principles such as ray tracing may be employed. These methods include geometrical optics (GO), the geometrical theory of diffraction (GTD) [4], and the physical theory of diffraction (PTD) [5]. In these cases accurate results are typically obtained only in limited regions of space. The accuracy of these methods improves as the size of the body increases [6].

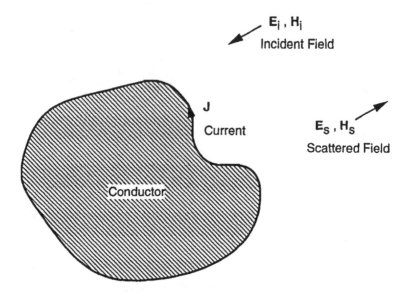

Figure 1.3: Scattering by a perfectly conducting body.

1.3   Scattering by Coated Conducting Bodies

Consider next the problem of scattering by a conducting body with a complex coating, as depicted in Figure 1.4. The field outside of the body is governed by Maxwell's equations for free space, the same set of equations as the field of the previous example. The field inside the coating is governed by a set of equations that take into account the detailed electromagnetic properties of the coating. These equations may be only slightly more complicated than those for free space, as in the case of a homogeneous dielectric coating, or may be quite complicated, as in the case of an inhomogeneous anisotropic material.

The first of two methods of attacking the exact solution of this problem is depicted in Figure 1.5. This method is essentially an extension of the method applied to the conducting scatterer and is applicable to homogeneous coatings. The tangential electric and magnetic fields at the boundary between free space and the exterior surface of the coating, as well as the electric current on the inner conductor, are considered as the unknowns [7]. The first condition to be enforced is that the tangential electric and magnetic fields be continuous across a boundary. In addition, the tangential electric field must be forced to vanish on the inner conductor. If the appropriate Green's functions for the coating material exist, a set of integral equations relating all of the unknown quantities may be obtained through the use of Green's theorem. For complex coating materials the Green's function may be quite complicated. The numerical solution of the problem is therefore much more computationally intensive than that for the conducting case. In addition, the number of unknowns is increased

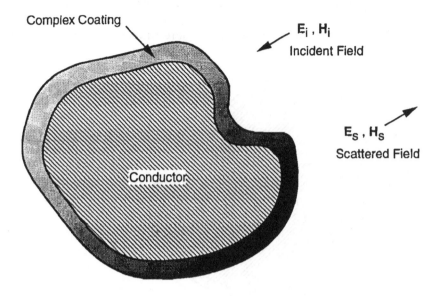

Figure 1.4: A conducting body with a complex coating.

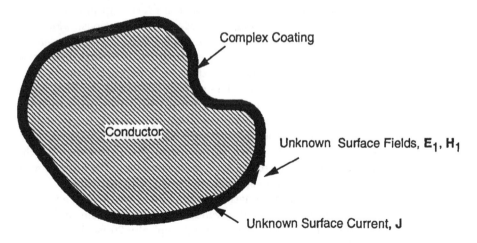

Figure 1.5: Equivalent current-based solution for homogeneous coatings.

relative to that of the conducting case. For a single thin coating the number of unknowns increases to 600 per square wavelength, and even small bodies require a formidable amount of computation. When more layers are present an additional number of unknowns, 400 per square wavelength, are required per layer.

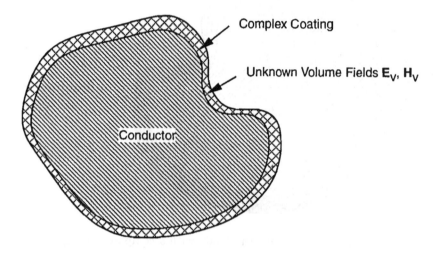

Figure 1.6: Finite element solution for inhomogeneous coatings.

The second approach involves solving for the fields at every point inside the volume of the coating and is therefore applicable to inhomogeneous coatings. In the finite element method (FEM) the coating is broken up into a number of small cells [8], as depicted in Figure 1.6. Maxwell's equations, which are partial differential equations, are solved inside the coating. The finite element mesh is truncated at the coating to free space boundary. An integral equation relating the fields on this boundary is obtained through the use of the free space Green's function and Green's theorem. The partial differential equation inside the coating, as well as the integral equation on the boundary, must then be solved simultaneously. An alternate approximation approach is to extend the finite element mesh outside the actual scatterer and employ an approximate absorbing boundary condition at the mesh boundary [9]. This eliminates the need for integral equations at the expense of an increased number of unknowns and obtaining only an approximate solution to the scattering problem.

The solution of the differential equation results in a sparse matrix equation, which may be solved efficiently. The disadvantage of the finite element method is that a large number of unknowns are required, typically 20 per linear wavelength, or $20^3 = 8000$ per cubic wavelength. To date the finite element method has only been applied to a limited number of three-dimensional geometries containing simple dielectric and conducting materials. Alternative formulations, using equivalent volume currents to model the presence of the coating material along with integral equations, also exist, but require approximately the same number of unknowns as the finite element method. The requirement of a large number of unknowns to represent the problem is the principal disadvantage of the finite element and related methods.

1.4 Approximate Boundary Conditions

With the previous discussion as background it is appropriate to discuss the topic of approximate boundary conditions, the subject of this book. The use of approximate boundary conditions in electromagnetic scattering has been an area of research for at least 40 years. Approximate boundary conditions may be used to eliminate many of the difficulties associated with the computation of scattering by coated bodies. Since the boundary conditions themselves are approximate, the difficulties are eliminated at the expense of obtaining only an approximate solution of the scattering problem. In many cases acceptable accuracy may be attained, and the approximate boundary conditions may be used to obtain solutions to scattering problems that cannot be solved using other methods.

The concept of approximate boundary conditions is based on the postulate that the relationship between the tangential electric and magnetic fields at any point on the boundary between the exterior of the coated body and free space is a purely local one, depending only on the properties of the coating directly below the point in question (see Figure 1.7). Given this relationship, the Green's function for free space, and Green's theorem, a solution of the scattering problem may be obtained. Once the relationship between the fields

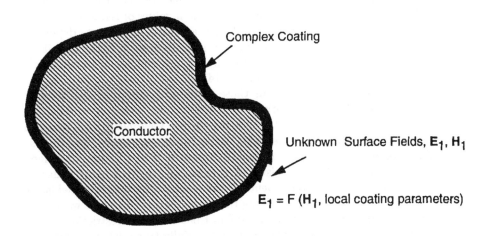

Figure 1.7: Solution using approximate boundary conditions for scattering by a coated body.

on the exterior of the coated body is known it is not necessary to consider the fields inside the coating or on the conductor as unknowns. The only true unknown is either the tangential electric or magnetic field on the free space boundary, the other being obtained through the use of the boundary condition. Thus, the number of unknowns for the multilayer coating problem becomes identical to that of the conducting body problem.

When the approximate boundary conditions are employed a considerable amount of computation time is saved. Assuming the approximate boundary condition accurately models the electromagnetic behavior of the coating, the loss in accuracy relative to the exact solution can be minimal. The accuracy of the approximate boundary condition is in turn determined by its complexity. Several existing approximate boundary condition theories will be described in the next few paragraphs, and the relationship between complexity and accuracy will become clear.

1.5   A Bit of History

The simplest and most widely used approximate boundary condition is the Standard Impedance Boundary Condition (SIBC) [10-13]. In this case the tangential components of the electric field are related to those of the magnetic fields by a simple multiplication factor. The multiplication factor is obtained by solving the problem reflection of a normally incident plane wave at an infinite ground plane coated with the appropriate material. Although simple in concept, this approximate boundary condition is quite accurate for some situations. In particular, when the reflection characteristics of the boundary are essentially the same for all angles of incidence the SIBC is an excellent approximate boundary condition. These conditions are met by very thin coatings, as well as by coatings made up of materials with high values of refractive index or significant loss. For coatings with reflection characteristics strongly dependent on the angle of incidence, the SIBC fails to give accurate results. Despite its limitations, the SIBC has been employed by a large number of researchers in the solution of several practical problems [14-16].

Many complex coatings, such as those containing nonreciprocal or anisotropic materials, rotate the polarization of the incident fields upon reflection. Polarization rotation also occurs with some surface treatments such as corrugated conductors [17]. For bodies with such coatings or surface treatments a boundary condition analogous to the SIBC, known as the Tensor Impedance Boundary Condition (TIBC), exists. It consists of a pair of linear equations that relate the tangential components of the electric and magnetic fields. As with the SIBC it is obtained by solving the normally incident plane wave problem for the boundary of interest. It suffers from the same restrictions as the SIBC. The TIBC has recently been used successfully to model the behavior of frequency selective surfaces at low frequencies [18].

Both the SIBC and the TIBC suffer from the same restriction: the behavior of the coating or surface treatment must be independent of the angle of incidence. This restriction is severe, essentially limiting their applicability to electrically thin or highly lossy coatings. By incorporating derivatives of the field components in the boundary condition the restriction that the properties of the coating or surface treatment not be a function of the angle of incidence is removed. Approximate boundary conditions that include derivatives of the field components are referred to as Higher Order Impedance Boundary Conditions (HOIBC). Such boundary conditions were considered in Ref. [19] in the 1960s, and later in Ref. [20].

More recently a set of boundary conditions based on the form of those discussed in Ref. [19], denoted as the Generalized Impedance Boundary Conditions (GIBC), have received a great deal of attention [21]. They relate the normal components of the electric and magnetic fields through a pair of differential equations. By including the normal derivatives of the normal field components in the boundary condition, one may simulate the angular dependence of the reflection characteristics of the boundary. This is precisely the effect that is ignored in the SIBC, and hence the GIBC are capable of simulating a wider range of coatings than the SIBC. The coefficients appearing in the differential equations depend on the local parameters of the coating and may be determined in a number of ways. They may be determined by examining the reflection characteristics of the planar coating for various angles of incidence [21] or by employing Taylor series expansions of the fields in the various layers making up the coating, [20, 22, 23].

Through the use of Maxwell's equations the normal derivatives and normal components of the fields may be written in terms of the tangential derivatives and fields. In this manner a pair of differential equations relating the tangential components of the fields and their tangential derivatives is obtained. The increased accuracy of the HOIBC is obtained at the expense of complexity. When the HOIBC is employed in the solution of a scattering problem differential equations as well as integral equations must be solved simultaneously, as opposed to simple linear equations and integral equations when the SIBC or TIBC is employed. When the conducting backing for the coating is absent the HOIBC are referred to as transition conditions [22], relating the tangential components of the fields and their derivatives on opposite sides of the thin layer. The GIBC have been used to solve a number of practical problems such as the canonical problem of scattering by a metal-backed dielectric half plane [24], a thin dielectric layer [25], a dielectric-filled groove in a ground plane [26], and two-dimensional coated bodies [27].

1.6 Organization of This Book

The first five chapters of this book develop a general theory of higher order impedance boundary conditions based on a spectral domain approach. The appropriate higher order impedance boundary conditions for a particular planar boundary are derived through consideration of the exact spectral domain solution to an appropriate canonical problem. The new approach is quite general, containing all of the existing approximate boundary condition theories, the SIBC, TIBC, and GIBC, as special cases. In addition, it is applicable to nonreciprocal and anisotropic coatings, as well as to surface treatments such as corrugations, which cannot be handled easily using the existing theories.

The new approach is based on approximating the exact spectral domain impedances in terms of a ratio of polynomials in the transform variables. This leads to a pair of differential equations relating the tangential electric and magnetic fields in the spatial domain. This approach eliminates the additional complicated steps of transferring the normal components and derivatives into tangential components and derivatives, a process that must be carried out before the GIBC may be applied to any practical problem. Using the new approach multilayer coatings may be handled in a straightforward manner because well-known transmission line techniques may be used to determine the exact boundary condition before it is approximated. Unlike the existing theories, the new approach can easily incorporate the effects of curvature into the boundary condition by considering the canonical problem of scattering by a cylinder rather than scattering by a plane, as will be shown.

After the general theory is described higher order boundary conditions will be derived for several planar and curved coatings. Their accuracy will then be determined by reapplying them to the canonical problems of scattering by coated conducting planes or coated conducting circular cylinders and comparing them to exact solutions for these geometries. The canonical problems of interest for this study are depicted in Figure 1.8.

As a first application of the theory the important special case of a planar dielectric layer, Figure 1.8a, will be discussed. The range of validity for the second order boundary condition will be discussed in detail. The effects of dielectric constant and coating depth on the accuracy of the HOIBC will be discussed, as well as the ability of the HOIBC to predict the correct surface wave behavior for the layer.

Next an example of a surface treatment as opposed to a coating will be considered. The corrugated conductor, Figure 1.8b, discussed earlier is chosen as an example. A straightforward application of the GIBC to this problem is not possible. However, this surface treatment can be considered as a simple extension to the planar dielectric layer when the spectral domain theory is employed. Next a coating made of chiral material [28], Figure 1.8c, will be considered. This material is chosen as an example since application of the SIBC to a coating containing this material is totally inadequate. Interesting effects occur for a chiral layer when it is illuminated at oblique angles of incidence, and only a higher order impedance boundary condition is capable of simulating these effects. Application of the GIBC to this problem is tedious, whereas the spectral domain approach requires no special modification to handle this more complex material. It will be shown that for many choices of coating parameters

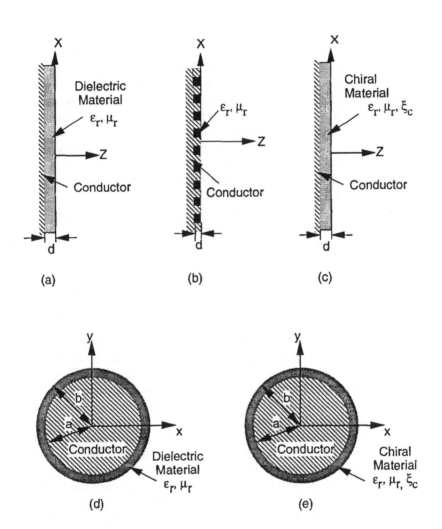

Figure 1.8: Canonical problems used to derive higher order impedance boundary conditions: (a) planar dielectric layer, (b) corrugated conductor, (c) planar chiral layer, (d) dielectric-coated circular cylinder, and (e) chiral-coated circular cylinder.

the second order HOIBC is capable of simulating the properties of this complex material with high accuracy.

Chapter 4 of the book discusses the effect of curvature on the accuracy of the planar based HOIBC. If the radii of curvature on the surface of the body are not much larger than the wavelength of interest, the planar approximation can introduce significant error. This is demonstrated by applying the planar HOIBC developed for dielectric and chiral coatings to the problem of scattering by two-dimensional circular cylinders coated with these materials, Figure 1.8d-e. It is shown that by considering the canonical problem of the coated circular cylinder rather than the coated ground plane, boundary conditions that account for finite curvature in one dimension may be derived. Once this canonical problem has been identified, an exact boundary condition is derived and approximated to derive the HOIBC in a manner identical to the planar solution discussed above. In contrast, existing theories such as the GIBC cannot be extended easily to include the effects of curvature.

Efficient numerical implementation of the approximate boundary condition theory with practical problems is a prime consideration. Chapters 5-7 of the book deal with application of the HOIBC to several practical problems. The specific problems of interest are depicted in Figure 1.9.

As a first example the problem of scattering by an arbitrarily shaped groove in a ground plane filled with dielectric layers is considered (see Figure 1.9a). The effects of groove depth, discontinuities in the groove depth, and the dielectric constants of the materials involved on the accuracy of the HOIBC solution are investigated. For the first time edge conditions will be employed along with the higher order boundary conditions in order to handle scattering by grooves with sharp corners. Results from the HOIBC and SIBC solutions are compared to those based on an exact formulation of the problem. Since this problem is completely planar, curvature effects are not an issue in this case, and they are considered in the following two examples.

The case of scattering by two-dimensional dielectric-coated cylinders will be considered in Chapter 6 (see Figure 1.9b). For the first time higher order impedance boundary conditions that include the effects of curvature will be applied. The accuracy of the higher order impedance boundary conditions that consider the effects of curvature are compared to those which are based on the planar approximation. Bistatic radar cross sections as well as monostatic radar cross sections are computed for a number of different shapes and coating parameters, and they are compared to those obtained from a method of moments solution based upon an exact formulation of the problem as well as the SIBC solution.

Previous researchers have focused on the application of higher order impedance boundary conditions to one- and two-dimensional problems. In the final example higher order impedance boundary conditions are applied to a three-dimensional scattering problem for the first time. The case of scattering by dielectric-coated conducting bodies of revolution (BOR) is addressed

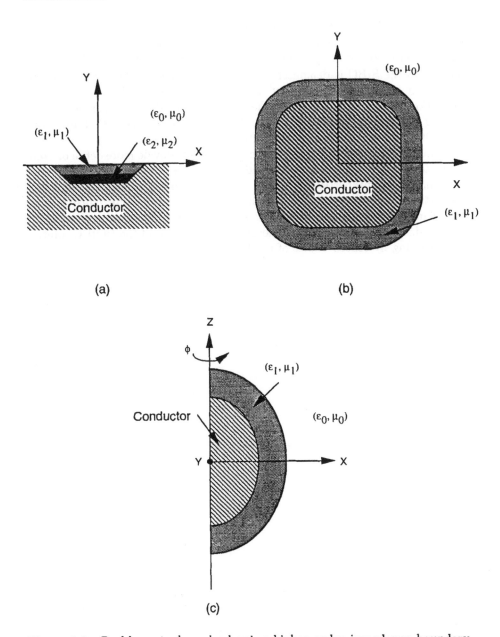

Figure 1.9: Problems to be solved using higher order impedance boundary conditions: (a) scattering by a dielectric-filled groove in a ground plane, (b) scattering by a two-dimensional dielectric-coated conducting cylinder, and (c) scattering by a three-dimensional dielectric-coated conducting body of revolution.

(see Figure 1.9c). Solutions for bistatic and monostatic RCS based on a planar HOIBC solution are compared to those using the SIBC and to an exact formulation of the problem. The increased accuracy of the HOIBC solution relative to the SIBC is demonstrated. The particular problem of transferring the planar higher order boundary conditions onto the doubly curved body of revolution is discussed in detail. The examples include a number of different coating parameters and BOR shapes. Limitations of the HOIBC imposed by the finite radii of curvature on the BOR are investigated.

A number of appendices are included in order to provide more detail on some of the topics addressed in the main body of the book. Derivations are provided for some of the properties of general impedance matrices and for symmetry properties of the boundary conditions. Surface waves on impedance surfaces and plane wave scattering by impedance planes and impedance cylinders are discussed. Several appendices are devoted to the details of the HOIBC solution for scattering by coated bodies of revolution.

Chapter 2

Spectral Domain Theory of Higher Order Impedance Boundary Conditions

In this chapter the topic of higher order impedance boundary conditions for planar coatings will be discussed in terms of a general spectral domain approach. Existing boundary condition methods such as the Standard Impedance Boundary Conditions (SIBC), Tensor Impedance Boundary Conditions (TIBC), and the Generalized Impedance Boundary Conditions (GIBC) are shown to be included in the theory as special cases.

In contrast to the existing methods for obtaining approximate boundary conditions, the spectral approach is applicable to a planar coating of material with arbitrary constitutive relations. In this respect the present approach is more versatile than the Generalized Impedance Boundary Conditions, GIBC. While the traditional approximate boundary conditions are based on the normal components of the field, applications typically require relationships between the tangential electric and magnetic fields. The spectral approach is based directly upon the transverse components of the magnetic and electric fields. In this respect the spectral domain approach is simpler than the GIBC, requiring fewer mathematical manipulations to obtain the boundary condition. In addition, it will be seen that it is relatively simple matter to include the effects of curvature using the new approach.

The four key steps in deriving the higher order impedance boundary conditions using the spectral domain approach are described in detail in this chapter. First, an exact tensor impedance boundary condition at the boundary between a conductor-backed layer of arbitrary material and free space is derived for each possible plane wave component. Second, an approximate spectral domain boundary condition is obtained by approximating the exact spectral domain impedances as ratios of polynomials in the transverse wavenumbers. Third, the unknown coefficients appearing in the polynomial approximation are determined. Finally, this approximate spectral domain boundary condition is then transferred into the spatial domain using the Fourier transform, resulting in a pair of differential equations relating the tangential electric and magnetic fields. Once the boundary condition is written in the spatial domain, it may be applied to nonplanar surfaces and to surfaces with coating parameters that vary from point to point, provided that the surface may be approximated as locally planar. The requirement that the surface be locally planar will be removed in Chapter 4, where higher order impedance boundary conditions that include the effects of curvature will be derived.

2.1 Step 1: Exact Spectral Domain Boundary Conditions

To derive a general set of higher order impedance boundary conditions in the spatial domain, that are applicable to relatively flat regions, the spectral domain problem of plane wave propagation in a conductor-backed layer of arbitrary material is considered. The pertinent geometry for the problem is depicted in Figure 2.1.

For each possible set of transverse wavenumbers (k_x, k_y), the total field inside the layer of material of thickness d may be written as a sum of four waves with unknown weighting coefficients $c_1, c_2, c_3,$ and c_4,

$$\mathbf{E}\left(k_x, k_y, z\right) = \sum_{i=1}^{4} c_i \mathbf{e}_i \left(k_x, k_y\right) e^{-j\left(k_x x + k_y y + k_z^{(i)} z\right)} \qquad (2.1)$$

and

$$\mathbf{H}\left(k_x, k_y, z\right) = \sum_{i=1}^{4} c_i \mathbf{h}_i \left(k_x, k_y\right) e^{-j\left(k_x x + k_y y + k_z^{(i)} z\right)}. \qquad (2.2)$$

The transverse vector wave functions $\mathbf{e}_i\left(k_x, k_y\right)$ and $\mathbf{h}_i\left(k_x, k_y\right)$, as well as the associated propagation constants $k_z^{(i)}$, are functions of the material properties, and can always be determined, even for materials with arbitrary constitutive relations [29]. For a simple dielectric layer $\mathbf{e}_i$ and $\mathbf{h}_i$ may be expressed in terms of TE to z and TM to z waves propagating in the positive and negative

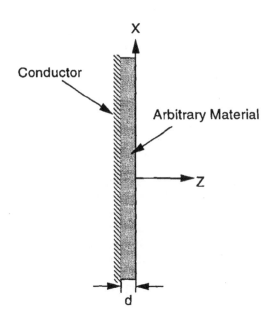

Figure 2.1: A conducting ground plane coated with a layer of arbitrary material.

z directions. As will be seen later, for a chiral layer the appropriate waves are right-hand and left-hand circularly polarized waves propagating in the positive and negative z directions. The above representation is valid for anisotropic and nonreciprocal coatings as well [30]. In this case the wavenumbers and wave functions are complicated functions of the constitutive relations for the coating.

For each set of transverse wavenumbers (k_x, k_y), a boundary condition at the interface, $z = 0$, may be obtained by eliminating the four unknown coefficients as follows. If the unknown coefficients are separated into pairs, then the tangential fields may be written in matrix form as

$$\{E(z)\} = [E_{12}(z)]\{C_{12}\} + [E_{34}(z)]\{C_{34}\}, \tag{2.3}$$

and

$$\{H(z)\} = [H_{12}(z)]\{C_{12}\} + [H_{34}(z)]\{C_{34}\}, \tag{2.4}$$

with

$$\{E(z)\} = \left\{ \begin{array}{c} E_x(z) \\ E_y(z) \end{array} \right\}, \ \{H(z)\} = \left\{ \begin{array}{c} H_x(z) \\ H_y(z) \end{array} \right\}, \tag{2.5}$$

$$\{C_{12}\} = \left\{ \begin{array}{c} c_1 \\ c_2 \end{array} \right\}, \ \{C_{34}\} = \left\{ \begin{array}{c} c_3 \\ c_4 \end{array} \right\}, \tag{2.6}$$

$$[E_{12}(z)] = \left[\begin{array}{cc} \hat{\mathbf{x}} \cdot \mathbf{e}_1 e^{-jk_z^{(1)}z} & \hat{\mathbf{x}} \cdot \mathbf{e}_2 e^{-jk_z^{(2)}z} \\ \hat{\mathbf{y}} \cdot \mathbf{e}_1 e^{-jk_z^{(1)}z} & \hat{\mathbf{y}} \cdot \mathbf{e}_2 e^{-jk_z^{(2)}z} \end{array} \right], \tag{2.7}$$

$$[E_{34}(z)] = \begin{bmatrix} \hat{\mathbf{x}} \cdot \mathbf{e}_3 e^{-jk_z^{(3)}z} & \hat{\mathbf{x}} \cdot \mathbf{e}_4 e^{-jk_z^{(4)}z} \\ \hat{\mathbf{y}} \cdot \mathbf{e}_3 e^{-jk_z^{(3)}z} & \hat{\mathbf{y}} \cdot \mathbf{e}_4 e^{-jk_z^{(4)}z} \end{bmatrix}, \tag{2.8}$$

$$[H_{12}(z)] = \begin{bmatrix} \hat{\mathbf{x}} \cdot \mathbf{h}_1 e^{-jk_z^{(1)}z} & \hat{\mathbf{x}} \cdot \mathbf{h}_2 e^{-jk_z^{(2)}z} \\ \hat{\mathbf{y}} \cdot \mathbf{h}_1 e^{-jk_z^{(1)}z} & \hat{\mathbf{y}} \cdot \mathbf{h}_2 e^{-jk_z^{(2)}z} \end{bmatrix}, \tag{2.9}$$

and

$$[H_{34}(z)] = \begin{bmatrix} \hat{\mathbf{x}} \cdot \mathbf{h}_3 e^{-jk_z^{(3)}z} & \hat{\mathbf{x}} \cdot \mathbf{h}_4 e^{-jk_z^{(4)}z} \\ \hat{\mathbf{y}} \cdot \mathbf{h}_3 e^{-jk_z^{(3)}z} & \hat{\mathbf{y}} \cdot \mathbf{h}_4 e^{-jk_z^{(4)}z} \end{bmatrix}, \tag{2.10}$$

where the $e^{-j(k_x x + k_y y)}$ dependence has been suppressed. By satisfying the boundary condition on the perfect conductor, $\{E(-d)\} = \{0\}$, the coefficients $\{C_{34}\}$ may be determined in terms of $\{C_{12}\}$. The remaining coefficients $\{C_{12}\}$ may be written in terms of the tangential components of the total magnetic field at the coating to air interface, $\{H(0)\}$. Finally, the two tangential components of the electric field at $z = 0$ may be written in terms of the two magnetic field components at $z = 0$,

$$\{E(0)\} = [Z]\{H(0)\}, \tag{2.11}$$

where

$$[Z] = \Big([E_{12}(0)] - [E_{34}(0)][E_{34}(-d)]^{-1}[E_{12}(-d)] \Big)$$

$$\Big([H_{12}(0)] - [H_{34}(0)][E_{34}(-d)]^{-1}[E_{12}(-d)] \Big)^{-1}. \tag{2.12}$$

Reinserting the dependence on the transverse wavenumbers explicitly, and using the tilde to indicate that the fields are evaluated at $z = 0$, the exact boundary condition in the spectral domain may be written as follows:

$$\begin{bmatrix} \tilde{E}_x(k_x, k_y) \\ \tilde{E}_y(k_x, k_y) \end{bmatrix} = \begin{bmatrix} Z_{xx}(k_x, k_y) & Z_{xy}(k_x, k_y) \\ Z_{yx}(k_x, k_y) & Z_{yy}(k_x, k_y) \end{bmatrix} \begin{bmatrix} \tilde{H}_x(k_x, k_y) \\ \tilde{H}_y(k_x, k_y) \end{bmatrix}, \tag{2.13}$$

where the terms of the impedance tensor are computed using Eq. (2.12). It should be noted that no incident field has been assumed and therefore the above relationship is purely a function of the coating parameters. It will be referred to as the *exact* spectral domain boundary condition. The notion "spectral domain" is used since Eq. (2.13) is held for each plane wave component with spectral parameter (k_x, k_y). In general the four entries in the impedance matrix (tensor) are not independent. Several general properties of this impedance tensor are derived in Appendix A. For arbitrary illumination of the coating, the fields may always be expressed as an infinite number of spectral domain components through the use of the Fourier transform. The above equation therefore describes a tensor impedance relationship between the Fourier transforms of the tangential electric and magnetic field components on the boundary.

In order to obtain the corresponding exact spatial domain boundary conditions it would be necessary to transform the exact spectral domain boundary condition equations into the spatial domain using the inverse Fourier transform. Due to the complicated nature of the four impedance functions $Z_{xx}(k_x, k_y)$, $Z_{xy}(k_x, k_y)$, $Z_{yx}(k_x, k_y)$, and $Z_{yy}(k_x, k_y)$, this is generally impossible unless some approximation to the impedance tensor is made. This is the subject of the following section. The development will continue assuming that the impedance terms are arbitrary functions, although specific forms of the impedance terms will be given for several coatings and surface treatments in the following chapter.

2.2   Step 2: Polynomial Approximations

A suitable higher order impedance boundary condition in the spatial domain may be obtained by applying the Fourier transform to the spectral domain boundary condition. It may be noted that polynomial expressions in the transform variables, k_x, and k_y may be transformed into differential equations in the spatial domain using elementary properties of the Fourier transform,

$$k_x \rightarrow j\frac{\partial}{\partial x}, \text{ and } k_y \rightarrow j\frac{\partial}{\partial y}. \tag{2.14}$$

The exact spectral domain boundary condition, Eq. (2.13), is therefore approximated by the following approximate boundary condition in the spectral domain,

$$
\begin{aligned}
&\left[\begin{array}{cc} P_1(k_x, k_y) & P_2(k_x, k_y) \\ P_3(k_x, k_y) & P_4(k_x, k_y) \end{array}\right]\left[\begin{array}{c} \tilde{E}_x(k_x, k_y) \\ \tilde{E}_y(k_x, k_y) \end{array}\right] \\
&\approx \left[\begin{array}{cc} P_5(k_x, k_y) & P_6(k_x, k_y) \\ P_7(k_x, k_y) & P_8(k_x, k_y) \end{array}\right]\left[\begin{array}{c} \tilde{H}_x(k_x, k_y) \\ \tilde{H}_y(k_x, k_y) \end{array}\right].
\end{aligned} \tag{2.15}
$$

The functions $P_1(k_x, k_y)$-$P_8(k_x, k_y)$ are polynomials in the transform variables, and for the second order case they take on the form

$$P_n(k_x, k_y) = c_0^n + c_1^n k_x + c_2^n k_y + c_3^n k_x^2 + c_4^n k_y^2 + c_5^n k_x k_y, \tag{2.16}$$

with c_m^n a complex number that is independent of k_x and k_y. The approximate boundary condition, Eq. (2.15), is easily transformed into a pair of differential equations that can be used to relate the spatial tangential components of the fields at each point on the exterior surface of the body. The electromagnetic behavior of the coating is then assumed to be represented entirely by the differential equations. Higher order polynomial approximations in the spectral domain will be more accurate, but will result in differential equations containing spatial derivatives of higher than second order. For the remainder of this book the discussion will center around the important special case of the second order polynomial approximation given above.

The exact spectral domain boundary condition and the approximate one are related to each other through the following equation,

$$
\begin{bmatrix} P_1(k_x,k_y) & P_2(k_x,k_y) \\ P_3(k_x,k_y) & P_4(k_x,k_y) \end{bmatrix} \begin{bmatrix} Z_{xx}(k_x,k_y) & Z_{xy}(k_x,k_y) \\ Z_{yx}(k_x,k_y) & Z_{yy}(k_x,k_y) \end{bmatrix}
$$
$$
\approx \begin{bmatrix} P_5(k_x,k_y) & P_6(k_x,k_y) \\ P_7(k_x,k_y) & P_8(k_x,k_y) \end{bmatrix}, \tag{2.17}
$$

or equivalently,

$$
\begin{bmatrix} Z_{xx}(k_x,k_y) & Z_{xy}(k_x,k_y) \\ Z_{yx}(k_x,k_y) & Z_{yy}(k_x,k_y) \end{bmatrix}
$$
$$
\approx \begin{bmatrix} P_1(k_x,k_y) & P_2(k_x,k_y) \\ P_3(k_x,k_y) & P_4(k_x,k_y) \end{bmatrix}^{-1} \begin{bmatrix} P_5(k_x,k_y) & P_6(k_x,k_y) \\ P_7(k_x,k_y) & P_8(k_x,k_y) \end{bmatrix}. \tag{2.18}
$$

The second relationship, Eq. (2.18), indicates that, using the present formulation, the terms in the impedance tensor are approximated as *ratios of polynomials* in the transform variables. The first equation, Eq. (2.17), is most convenient for determining the 48 unknown coefficients. A simple method for determining the unknown coefficients is described in the next section.

2.3   Step 3: Coefficient Determination

The next step is to determine the 48 coefficients appearing in the polynomials. In many cases certain symmetries of the coating can be used to eliminate many of the coefficients. A general method for considering the effects of rotational symmetry on the boundary condition is discussed in Appendix B. For example, if the coating is invariant under a 180 degree rotation it can be shown that the linear terms must vanish, i.e., $c_1^n = c_2^n = 0$. In a great number of practical situations the boundary condition must be invariant under an arbitrary rotation of the x-y coordinate system. Examples of this include planar dielectric layers, chiral layers, normally biased ferrite layers, and crystals oriented so that the axis of symmetry is normal to the x-y plane. In this case it can be shown through symmetry that the polynomials take on the simpler forms,

$$
P_1(k_x,k_y) = 1 + c_3^1 k_x^2 + c_4^1 k_y^2 + (c_4^2 - c_3^2)k_x k_y, \quad P_4(k_x,k_y) = P_1(k_y,-k_x), \tag{2.19}
$$

$$
P_2(k_x,k_y) = c_3^2 k_x^2 + c_4^2 k_y^2 + (c_3^1 - c_4^1)k_x k_y, \quad P_3(k_x,k_y) = -P_2(k_y,-k_x), \tag{2.20}
$$

$$
P_5(k_x,k_y) = c_0^5 + c_3^5 k_x^2 + c_4^5 k_y^2 + (c_4^6 - c_3^6), \quad P_8(k_x,k_y) = P_5(k_y,-k_x), \tag{2.21}
$$

and

$$P_6\left(k_x, k_y\right) = c_0^6 + c_3^6 k_x^2 + c_4^6 k_y^2 + \left(c_3^5 - c_4^5\right), \quad P_7\left(k_x, k_y\right) = -P_6\left(k_y, -k_x\right). \quad (2.22)$$

The coefficient c_0^1 has been set equal to 1 by virtue of the fact that the two boundary condition equations may always be scaled to make this possible. The coefficients c_0^2 and c_0^3 may be set to zero by virtue of the fact that linear combinations of the two equations may always be taken to force this condition.

Thus, for the rotationally invariant case there are ten unknown coefficients to be determined. Due to the symmetry of the situation it is also sufficient to examine the case $k_y = 0$. A general method for determining the ten coefficients is to recognize that the matrix equation, Eq. (2.17), is actually four separate equations. If the approximation is forced to satisfy the equation exactly for normal incidence and two distinct values of k_x, then there are exactly the correct number of equations to solve for the coefficients. When the approximate boundary condition does a good job of simulating the behavior of the coating the value of the resulting coefficients is insensitive to the choice of match points. One possible choice is to span the visible range for k_x, satisfying Eq. (2.17) for $k_x = 0$, $k_x = k_0/2$, and $k_x = k_0$, where k_0 is the free space wavenumber. This choice of match points produces the best fit in the visible range, but the fit must be acceptable in the surface wave region as well if the contribution from surface wave phenomena is to be predicted accurately. Any two distinct points may be chosen along with $k_x = 0$ in order to determine the coefficients. This will be discussed detail in the following chapters. Here we proceed assuming only that the two additional points are distinct.

Using the first match point $k_x = 0$, two of the coefficients are found immediately to be

$$c_0^5 = Z_{xx}(0,0) = Z_{yy}(0,0) \text{ and } c_0^6 = Z_{xy}(0,0) = -Z_{yx}(0,0). \quad (2.23)$$

The remaining eight coefficients are then found from matching the equations at $k_x^{(1)}$ and $k_x^{(2)}$. If

$$[M] = \begin{bmatrix} k_x^{(1)^2} Z_{xx}^{(1)} & k_x^{(1)^2} Z_{yx}^{(1)} & -k_x^{(1)^2} & 0 \\ k_x^{(1)^2} Z_{xy}^{(1)} & k_x^{(1)^2} Z_{yy}^{(1)} & 0 & -k_x^{(1)^2} \\ k_x^{(2)^2} Z_{xx}^{(2)} & k_x^{(2)^2} Z_{yx}^{(2)} & -k_x^{(2)^2} & 0 \\ k_x^{(2)^2} Z_{xy}^{(2)} & k_x^{(2)^2} Z_{yy}^{(2)} & 0 & -k_x^{(2)^2} \end{bmatrix}, \quad (2.24)$$

$$\{C_1\} = \begin{Bmatrix} c_3^1 \\ c_3^2 \\ c_3^5 \\ c_3^6 \end{Bmatrix}, \quad \{C_2\} = \begin{Bmatrix} -c_2^4 \\ c_4^1 \\ -c_4^6 \\ c_4^5 \end{Bmatrix}, \quad (2.25)$$

$$\{Z_1\} = \left\{ \begin{array}{c} Z_{xx}^{(0)} - Z_{xx}^{(1)} \\ Z_{xy}^{(0)} - Z_{xy}^{(1)} \\ Z_{xx}^{(0)} - Z_{xx}^{(2)} \\ Z_{xy}^{(0)} - Z_{xy}^{(2)} \end{array} \right\}, \text{ and } \{Z_2\} = \left\{ \begin{array}{c} Z_{yx}^{(0)} - Z_{yx}^{(1)} \\ Z_{yy}^{(0)} - Z_{yy}^{(1)} \\ Z_{yx}^{(0)} - Z_{yx}^{(2)} \\ Z_{yy}^{(0)} - Z_{yy}^{(2)} \end{array} \right\}, \tag{2.26}$$

then

$$[M]\{C_1\} = \{Z_1\} \text{ and } [M]\{C_2\} = \{Z_2\}, \tag{2.27}$$

with $Z_{\alpha\beta}^{(n)} = Z_{\alpha\beta}(k_x^{(n)}, 0)$ and $k_x^{(0)} = 0$.

Since the matrix $[M]$ appears in both equations, only a single 4×4 matrix decomposition needs to be performed to determine the coefficients. It can also be shown that this method for determining the coefficients satisfies reciprocity if the coating material has this property. As shown in Appendix A, reciprocity implies that the impedance tensor satisfies $Z_{xx}(k_x, k_y) = -Z_{yy}(k_x, k_y)$.

2.4   Step 4: Construction of the Spatial Domain Equations

Once the coefficients are determined it is a simple matter to transform the polynomial approximation of the spectral domain boundary condition into the spatial domain using elementary properties of the Fourier transform. The general form of the resulting pair of differential equations for the case of a rotational invariant coating are given below.

$$\begin{aligned}
&(1 - c_3^1 \frac{\partial^2}{\partial x^2} - c_4^1 \frac{\partial^2}{\partial y^2} - (c_4^2 - c_3^2)\frac{\partial^2}{\partial x \partial y})E_x(x,y) \\
&+(-c_3^2 \frac{\partial^2}{\partial x^2} - c_4^2 \frac{\partial^2}{\partial y^2} - (c_3^1 - c_4^1)\frac{\partial^2}{\partial x \partial y})E_y(x,y) \\
&= (c_0^5 - c_3^5 \frac{\partial^2}{\partial x^2} - c_4^5 \frac{\partial^2}{\partial y^2} - (c_4^6 - c_3^6)\frac{\partial^2}{\partial x \partial y})H_x(x,y) \\
&+(c_0^6 - c_3^6 \frac{\partial^2}{\partial x^2} - c_4^6 \frac{\partial^2}{\partial y^2} - (c_3^5 - c_4^5)\frac{\partial^2}{\partial x \partial y})H_y(x,y)
\end{aligned} \tag{2.28}$$

and

$$\begin{aligned}
&(c_4^2 \frac{\partial^2}{\partial x^2} + c_3^2 \frac{\partial^2}{\partial y^2} - (c_3^1 - c_4^1)\frac{\partial^2}{\partial x \partial y})E_x(x,y) \\
&+(1 - c_4^1 \frac{\partial^2}{\partial x^2} - c_3^1 \frac{\partial^2}{\partial y^2} + (c_4^2 - c_3^2)\frac{\partial^2}{\partial x \partial y})E_y(x,y) \\
&= (-c_0^6 + c_4^6 \frac{\partial^2}{\partial x^2} + c_3^6 \frac{\partial^2}{\partial y^2} - (c_3^5 - c_4^5)\frac{\partial^2}{\partial x \partial y})H_x(x,y) \\
&+(c_0^5 - c_4^5 \frac{\partial^2}{\partial x^2} - c_3^5 \frac{\partial^2}{\partial y^2} + (c_4^6 - c_3^6)\frac{\partial^2}{\partial x \partial y})H_y(x,y).
\end{aligned} \tag{2.29}$$

Table 2.1: Special cases of the higher order impedance boundary conditions.

Boundary Condition	Non-Zero Coefficients	Impedance Tensor
SIBC	c_0^6	$\begin{bmatrix} 0 & c_0^6 \\ -c_0^6 & 0 \end{bmatrix}$
TIBC	$c_0^5, c_0^6, c_0^7, c_0^8$	$\begin{bmatrix} c_0^5 & c_0^6 \\ c_0^7 & c_0^8 \end{bmatrix}$
3rd Order GIBC	$c_0^6, c_3^6, c_3^1, c_4^6, c_4^1$	$\begin{bmatrix} 0 & \frac{c_0^6 + c_3^6 k_x^2}{1 + c_3^1 k_x^2} \\ -\frac{c_0^6 + c_4^6 k_x^2}{1 + c_4^1 k_x^2} & 0 \end{bmatrix}$

These two differential equations are then postulated to describe the behavior of the coating as far as fields external to the coating ($z > 0$) are considered. They may be used on an arbitrarily shaped coated body on a point by point basis if the body may be considered to be locally planar. If the coating parameters vary from point to point then the coefficients are functions of position. As is discussed in Ref. [31], when the coating parameters do not have sufficient continuity additional constraints are required in order to obtain a unique solution to a given scattering problem. Aspects of this problem will be addressed in Chapter 5, where scattering by a groove in a ground plane is considered. It will be shown throughout the applications portion of this work that when the differential equations are employed in conjunction with the free space Green's function in the solution of a scattering problem, acceptable results may be obtained using significantly fewer unknowns than are required in an exact formulation of the problem.

2.5 Special Cases

In this section several special cases of the general theory are discussed. As is shown in Table 2.1, various choices of nonzero coefficients recover the SIBC, TIBC, and GIBC. These choices are described in detail in the next few paragraphs.

2.5.1   TIBC

It is interesting to note that if each of the four impedance terms is approximated by its value at normal incidence, i.e., $k_x = k_y = 0$, then $c_0^5 = Z_{xx}(0,0)$, $c_0^6 = Z_{xy}(0,0)$, $c_0^7 = Z_{yx}(0,0)$, and $c_0^8 = Z_{yy}(0,0)$, and the remaining coefficients vanish. The resulting boundary condition reduces to the Tensor Impedance Boundary Condition (TIBC) discussed in Ref. [18]. The boundary condition equations are no longer differential equations but simply a pair of constant coefficient equations relating the tangential field components. Such a boundary condition is accurate only when the impedances are essentially constant for all values of k_x and k_y. This implies that the waves inside the coating have large longitudinal wavenumbers and propagate nearly normal to the interface regardless of the transverse wavenumbers k_x and k_y.

2.5.2   SIBC

A further simplification occurs if there is no depolarization upon reflection for a normally incident wave, for example when the coating is a simple dielectric. In this case $Z_{xx}(0,0) = Z_{yy}(0,0) = 0$ and $Z_{xy}(0,0) = -Z_{yx}(0,0) = \eta$, and only c_0^6 is nonzero. The boundary condition then becomes the Standard Impedance Boundary Condition (SIBC), or the Leontovich boundary condition. This is the most widely used approximate boundary condition. It is has limited accuracy, and it is accurate only for thin layers with high indices of refraction or high loss.

2.5.3   GIBC

If $Z_{xx} = Z_{yy} = 0$ for all k_x when $k_y = 0$, such as for a dielectric coating, then the coefficients c_4^2, c_3^2, c_4^5, and c_3^5 vanish. Application of the Higher Order Impedance Boundary Condition (HOIBC) is then equivalent to approximating the two impedances $Z_{xy}(k_x, 0)$ and $Z_{yx}(k_x, 0)$ as ratios of polynomials in k_x,

$$Z_{xy}(k_x, 0) \approx \frac{c_0^6 + c_3^6 k_x^2}{1 + c_3^1 k_x^2} \text{ and } Z_{yx}(k_x, 0) \approx -\frac{c_0^6 + c_4^6 k_x^2}{1 + c_4^1 k_x^2}. \qquad (2.30)$$

The polynomial approximation then results in differential equations that have exactly the same form as the third order Generalized Impedance Boundary Conditions (GIBC) [21]. The present method for determining the coefficients appearing in the differential equations is, however, more general than that used in the GIBC. Unlike the GIBC the same procedure may be used to determine the coefficients for either low or high contrast coatings. This will be demonstrated in the next chapter, which applies the above theory to determine the HOIBC for a simple dielectric coating.

2.6   Conclusions

In this chapter a general spectral domain approach for deriving higher order impedance boundary conditions for planar coatings has been presented. It is applicable to multiple layers of arbitrary material. All of the existing boundary conditions, the SIBC, TIBC, and GIBC, are included in the approach as special cases. The general approach presented has wider applicability than the others and is also simpler, requiring fewer mathematical manipulations to obtain the boundary condition equations. As demonstrated, no special procedures are required for high or low contrast coatings. The following chapter will demonstrate how the general theory can be applied to simple dielectric coatings as well as two boundaries that are not easily handled using the existing boundary condition theories, chiral coatings, and corrugated conductors.

Chapter 3

Planar Higher Order Impedance Boundary Conditions

In this chapter the general theory described in Chapter 2 is applied to the several planar boundaries. The specific case of a dielectric layer on a conductor is considered first. These results are then extended to the case of a treated boundary, the corrugated conductor. Finally a more complex coating, a chiral layer on a conductor, is considered, demonstrating the applicability of the spectral domain approach to the determination of HOIBC for complex coatings. In each case the accuracy of the approximate boundary conditions as a function of the parameters of the coating or surface treatment is discussed.

3.1 Planar Dielectric Coatings

In this section the general theory presented in the previous chapter is applied to the case of a dielectric coating. Although simple, this particular coating is the most common and perhaps the most useful of all coatings. Lossy dielectric coatings can be employed to reduce the radar cross section of conducting bodies.

In addition, microstrip circuits and antennas are often printed upon a simple dielectric layer over a ground plane.

3.1.1 Exact Spectral Domain Boundary Conditions

The first step in the application of the theory discussed in Chapter 2 to dielectric coatings is to determine the four tangential electric and magnetic field vectors. As was mentioned earlier, for the dielectric layer these may be expressed as forward and reverse traveling TE to z and TM to z waves as

$$\mathbf{e}_1 = k_x\hat{\mathbf{x}} + k_y\hat{\mathbf{y}}, \quad \mathbf{h}_1 = -\frac{k}{k_z\eta}(k_y\hat{\mathbf{x}} - k_x\hat{\mathbf{y}}), \tag{3.1}$$

$$\mathbf{e}_2 = k_x\hat{\mathbf{x}} + k_y\hat{\mathbf{y}}, \quad \mathbf{h}_2 = \frac{k}{k_z\eta}(k_y\hat{\mathbf{x}} - k_x\hat{\mathbf{y}}), \tag{3.2}$$

$$\mathbf{e}_3 = k_y\hat{\mathbf{x}} - k_x\hat{\mathbf{y}}, \quad \mathbf{h}_3 = \frac{k_z}{k\eta}(k_x\hat{\mathbf{x}} - k_y\hat{\mathbf{y}}), \tag{3.3}$$

and

$$\mathbf{e}_4 = k_y\hat{\mathbf{x}} - k_x\hat{\mathbf{y}}, \quad \mathbf{h}_4 = -\frac{k_z}{k\eta}(k_x\hat{\mathbf{x}} - k_y\hat{\mathbf{y}}), \tag{3.4}$$

where $\eta = \sqrt{\mu/\epsilon}$, $k = \omega\sqrt{\mu\epsilon}$, and $k_z = \sqrt{k^2 - k_x^2 - k_y^2}$.

Following the methods described earlier in Chapter 2 the following exact impedances are obtained for the dielectric coated conductor,

$$Z_{xx}(k_x, k_y) = -Z_{yy}(k_x, k_y) = -j\sqrt{\frac{\mu}{\epsilon}}\frac{k_x k_y}{k k_z}\tan(k_z d), \tag{3.5}$$

$$Z_{xy}(k_x, k_y) = -Z_{yx}(k_y, k_x) = -j\sqrt{\frac{\mu}{\epsilon}}\frac{k_x^2 k_z^2 + k_y^2 k^2}{k k_z (k_x^2 + k_y^2)}\tan(k_z d), \tag{3.6}$$

with $\mu = \mu_r\mu_0$ and $\epsilon = \epsilon_r\epsilon_0$, which become complex quantities for a lossy coating.

The exact electromagnetic behavior of the coating, including any surface wave behavior, is contained in these impedance terms. For example, consider these impedance terms for a simple dielectric coating with $\epsilon_r = 4$, $\mu_r = 1.0$. For a thin layer it is well known that the Standard Impedance Boundary Condition is a good model, whereas for a thick coating the SIBC fails. The effect of coating thickness on the complexity of the impedance terms is illustrated in the next few figures. Figure 3.1 illustrates that for thin layers, in this case $d = 0.02\lambda_0$, the impedance term Z_{xy} is quite flat and essentially independent of (k_x, k_y), and thus the SIBC is an excellent approximation for the coating's behavior. The increasing complexity of the impedance terms as the coating becomes thicker is illustrated in Figure 3.2 and Figure 3.3. In the case of Figure 3.2, $d = 0.075\lambda_0$,

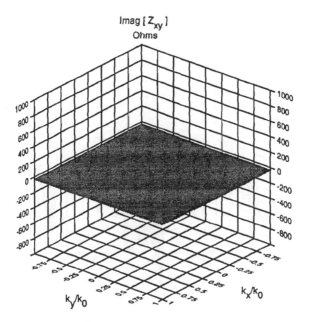

Figure 3.1: Imaginary part of Z_{xy} versus k_x and k_y for a dielectric coating with $\epsilon_r = 4.0$, $\mu_r = 1.0$, and $d = 0.02\lambda_0$.

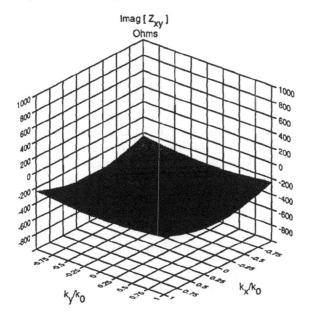

Figure 3.2: Imaginary part of Z_{xy} versus k_x and k_y for a dielectric coating with $\epsilon_r = 4.0$, $\mu_r = 1.0$, and $d = 0.075\lambda_0$.

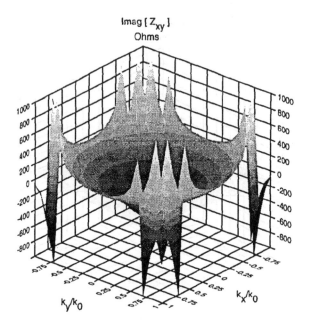

Figure 3.3: Imaginary part of Z_{xy} versus k_x and k_y for a dielectric coating with $\epsilon_r = 4.0$, $\mu_r = 1.0$, and $d = 0.2\lambda_0$.

and a second order ratio of polynomial approximation to the impedance is quite satisfactory; the HOIBC may be used to simulate the behavior of the coating. As illustrated in Figure 3.3, when $d = 0.2\lambda_0$ the behavior of the impedance terms is quite complex and the ratio of polynomial approximation offered by the second order HOIBC is found to be unsatisfactory. In this case derivatives of higher than second order must be included in the boundary condition in order to obtain an accurate model of the coating's behavior for the entire spectrum. This indicates that the behavior of the coating is becoming less localized, as expected for a thick coating. In the next section the effects illustrated in the previous three figures will be quantified, and a range of coating parameters that may be successfully simulated by a given higher order impedance boundary condition will be determined.

3.1.2 Higher Order Impedance Boundary Conditions

Three specific approximate boundary conditions for the planar dielectric layer will be discussed in detail in this section. All are based upon the ratio of polynomial expansion for the impedances discussed in Eq. (2.30),

$$Z_{xy}(k_x, 0) \approx \frac{c_0^6 + c_3^6 k_x^2}{1 + c_3^1 k_x^2} \text{ and } Z_{yx}(k_x, 0) \approx -\frac{c_0^6 + c_4^6 k_x^2}{1 + c_4^1 k_x^2}. \tag{3.7}$$

Recall that since the coating is invariant under rotation and $Z_{xx}(k_x, k_y) = Z_{yy}(k_x, k_y) = 0$ when $k_y = 0$ the complete spatial domain boundary condition may be determined in terms of the five unknown coefficients appearing above.

The simplest approximation is the SIBC, which sets $c_3^6 = c_3^1 = c_4^6 = c_4^1 = 0$ and c_0^6 equal to the impedance for normal incidence,

$$c_0^6 = -j\sqrt{\frac{\mu_0 \mu_r}{\epsilon_0 \epsilon_r}} \tan(\omega\sqrt{\mu_r \epsilon_r \mu_0 \epsilon_0}d). \tag{3.8}$$

As was discussed earlier, when this boundary condition is transformed into the spatial domain it is a simple linear equation involving no differential operations.

$$E_y(x, y) = -c_0^6 H_x(x, y), \tag{3.9}$$

and

$$E_x(x, y) = c_0^6 H_y(x, y). \tag{3.10}$$

The examples that follow will illustrate that the simplicity of the SIBC results in severe limitations on its applicability.

The accuracy of the boundary condition may be improved by including the coefficients c_3^6 and c_4^6, thus approximating the impedances by simple second order polynomials. In this case the coefficient c_0^6 is given by Eq. (3.8), and the remaining two coefficients are determined by matching the impedances at some arbitrary value of k_x, denoted by $k_x^{(1)}$,

$$c_3^6 = \frac{Z_{xy}(k_x^{(1)}, 0) - Z_{xy}(0, 0)}{k_x^{(1)2}}, \tag{3.11}$$

and

$$c_4^6 = -\frac{Z_{yx}(k_x^{(1)}, 0) - Z_{yx}(0, 0)}{k_x^{(1)2}}. \tag{3.12}$$

Upon transformation into the spatial domain this boundary condition results in a pair of differential equations that may be used to determine the tangential electric field in terms of the tangential magnetic field and its tangential derivatives,

$$E_y(x, y) = \left(-c_0^6 + c_4^6 \frac{\partial^2}{\partial x^2} + c_3^6 \frac{\partial^2}{\partial y^2}\right) H_x(x, y) + (c_4^6 - c_3^6)\frac{\partial^2 H_y(x, y)}{\partial x \partial y} \tag{3.13}$$

and

$$E_x(x, y) = (c_4^6 - c_3^6)\frac{\partial^2 H_x(x, y)}{\partial x \partial y}) + \left(c_0^6 - c_3^6 \frac{\partial^2}{\partial x^2} - c_4^6 \frac{\partial^2}{\partial y^2}\right) H_y(x, y). \tag{3.14}$$

Although this set of boundary conditions is more complicated than those of the SIBC, Eqs. (3.9) and (3.10), differential operators appear only for the magnetic field and thus, given the magnetic field, the electric field may be obtained immediately. It will be shown that Eqs. (3.13) and (3.14) are capable of simulating a wider range of coating parameters than the SIBC equations.

The final step in the process is to approximate the impedances as ratios of second order polynomials. In this case the five coefficients may be determined by matching the impedances exactly for normal incidence, resulting in Eq. (3.8), as well as at two unique values of k_x, $k_x^{(1)}$ and $k_x^{(2)}$, resulting in two pairs of linear equations for the remaining four coefficients,

$$Z_{xy}(k_x^{(1)},0) + c_3^1 k_x^{(1)^2} Z_{xy}(k_x^{(1)},0) - Z_{xy}(0,0) - c_3^6 k_x^{(1)^2} = 0 \tag{3.15}$$

and

$$Z_{xy}(k_x^{(2)},0) + c_3^1 k_x^{(2)^2} Z_{xy}(k_x^{(2)},0) - Z_{xy}(0,0) - c_3^6 k_x^{(2)^2} = 0, \tag{3.16}$$

$$Z_{yx}(k_x^{(1)},0) + c_4^1 k_x^{(1)^2} Z_{yx}(k_x^{(1)},0) - Z_{yx}(0,0) + c_4^6 k_x^{(1)^2} = 0 \tag{3.17}$$

and

$$Z_{yx}(k_x^{(2)},0) + c_4^1 k_x^{(2)^2} Z_{yx}(k_x^{(2)},0) - Z_{yx}(0,0) + c_4^6 k_x^{(2)^2} = 0. \tag{3.18}$$

The additional degree of freedom offered by the ratio of polynomial approximation of the impedances over the simple polynomial approximation greatly extends the useful range of the approximate boundary conditions. The spatial domain boundary conditions are now more complex, relating the tangential electric field as well as its tangential derivatives to the tangential magnetic field and its tangential derivatives,

$$\left(1 - c_3^1 \frac{\partial^2}{\partial x^2} - c_4^1 \frac{\partial^2}{\partial y^2}\right) E_x(x,y) + (c_3^1 - c_4^1)\frac{\partial^2 E_y(x,y)}{\partial x \partial y}$$
$$= (c_4^6 - c_3^6)\frac{\partial^2 H_x(x,y)}{\partial x \partial y} + \left(c_0^6 - c_3^6 \frac{\partial^2}{\partial x^2} - c_4^6 \frac{\partial^2}{\partial y^2}\right) H_y(x,y), \tag{3.19}$$

and

$$(c_3^1 - c_4^1)\frac{\partial^2 E_x(x,y)}{\partial x \partial y} + \left(1 - c_4^1 \frac{\partial^2}{\partial x^2} - c_3^1 \frac{\partial^2}{\partial y^2}\right) E_y(x,y)$$
$$= \left(-c_0^6 + c_4^6 \frac{\partial^2}{\partial x^2} + c_3^6 \frac{\partial^2}{\partial y^2}\right) H_x(x,y) + (c_4^6 - c_3^6)\frac{\partial^2 H_y(x,y)}{\partial x \partial y}. \tag{3.20}$$

In order to illustrate the relative accuracy of the three boundary conditions versus the parameters of the coating several examples will be presented in the next section.

3.1.3 Examples

Consider first a single-layer dielectric coating with $\epsilon_r = 4.0$, $\mu_r = 1.0$, and $d = 0.05\lambda_0$. The exact impedances as well as the three approximations are plotted in Figure 3.4. For the simple polynomial approximation $k_x^{(1)}$ has been

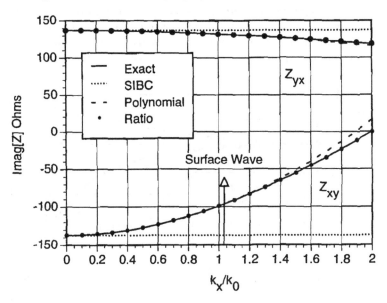

Figure 3.4: Impedances for a planar dielectric coating with $\epsilon_r = 4.0$, $\mu_r = 1.0$, and $d = 0.05\lambda_0$.

set equal to k_0, corresponding to a perfect match of the boundary condition for grazing incidence. For the ratio of polynomial approximation the additional match point $k_x^{(2)}$ is set equal to $2k_0$, spanning both the visible and surface wave regions for k_x. The impedances are plotted over the range $0 < k_x < 2k_0$, which includes the visible range of k_x, i.e., $0 < k_x < k_0$, as well as the range of k_x that corresponds to possible surface wave behavior, $k_0 < k_x < \sqrt{\epsilon_r \mu_r} k_0 = 2k_0$. The location of the surface wave wavenumbers for a boundary described by a general impedance tensor is discussed in Appendix C. In this particular case a single surface wave that is TM to the direction of propagation exists, as indicated on the figure. The ratio of polynomial approximation for the impedance is excellent, and the simple polynomial approximation is quite adequate. The SIBC is, however, a poor model for the behavior of the impedance. Note that although the match point, $k_x^{(1)}$, is chosen in the visible range of k_x, the simple polynomial approximation is adequate in the surface wave region as well. Thus, as discussed in Appendix C, the surface wave behavior of the layer is also well approximated by the HOIBC.

As an additional check on the accuracy of the approximate boundary conditions the problem of plane wave reflection from the coated ground plane is considered. The formulation of this problem in terms of the exact and approximate impedance tensors for the surface is outlined in Appendix D. While it is recognized that such a check does not verify the accuracy of the boundary condition in the surface wave region, it has been used previously to assess the accuracy of higher order impedance boundary conditions [21]. The reflection coefficients for plane wave incidence obtained from scattering by the exact layer, as well as those obtained using the various boundary conditions to simulate the layer, are shown in Figure 3.5. In all cases the magnitude of the

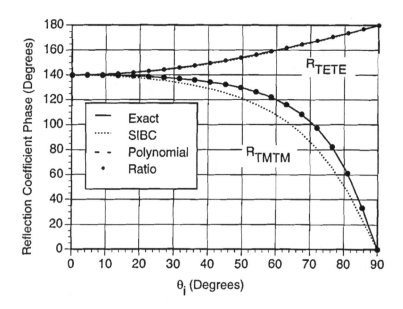

Figure 3.5: Phase of the reflection coefficients for a planar dielectric coating with $\epsilon_r = 4.0$, $\mu_r = 1.0$, and $d = 0.05\lambda_0$.

reflection coefficients is equal to unity, as required for a lossless layer. Since the approximate impedances are good approximations to the exact ones for both higher order boundary condition approximations, the excellent agreement demonstrated in Figure 3.5 is expected. The SIBC results are poor, particularly for TM polarization where the approximation for the pertinent impedance term Z_{xy} is poorest. The next example will demonstrate that as the coating becomes thicker the accuracy of the simple polynomial approximation decreases and a ratio of polynomials is required to simulate the coating over the allowable range of wavenumbers.

As a second example the coating thickness of the previous examples is increased to $d = 0.1\lambda_0$. The impedance and reflection coefficient results for this coating are presented in Figure 3.6 and Figure 3.7. For this thicker coating only

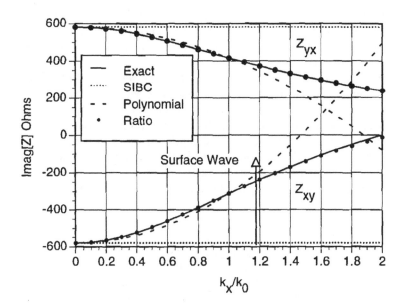

Figure 3.6: Impedances for a planar dielectric coating with $\epsilon_r = 4.0$, $\mu_r = 1.0$, and $d = 0.1\lambda_0$.

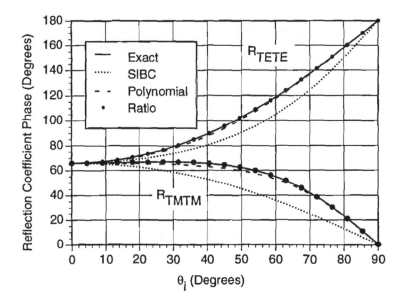

Figure 3.7: Phase of the reflection coefficients for a planar dielectric coating with $\epsilon_r = 4.0$, $\mu_r = 1.0$, and $d = 0.1\lambda_0$.

the accuracy of the simple polynomial approximation starts to deteriorate. On the other hand, Figure 3.6 shows that the ratio of polynomial approximation produces an acceptable result both in the visible range and in the surface wave region.

Altering the match points for the examples presented will obviously affect the accuracy of the various approximations, but several general observations can be made. In all cases the accuracy of the ratio of polynomial approximation exceeds that of the simple polynomial approximation, and for all but the thinnest coatings, in terms of wavelengths in the dielectric, the SIBC approximation is poor. The accuracy of the various approximations decreases for increasing coating depths and for increasing dielectric constants. These effects are depicted graphically in Figure 3.8 through Figure 3.10, where the average

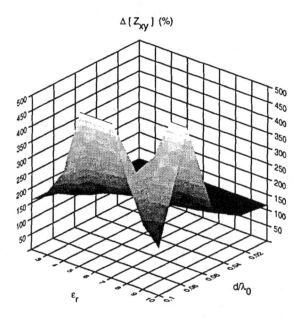

Figure 3.8: Plot of the average error between the SIBC and the exact imped-ance, Z_{xy}, for a range of ϵ_r and d, when $\mu_r = 1.0$.

error between the exact Z_{xy} and the various approximate impedances is plotted versus dielectric constant and coating thickness. The error measure ΔZ_{xy} has been chosen as

$$\Delta Z_{xy} = 100 \frac{\left| Z_{xy} - \tilde{Z}_{xy} \right|}{|Z_{xy}|}. \tag{3.21}$$

where $\tilde{Z}_{xy}$ is the approximate impedance and Z_{xy} is the exact impedance. The average error for k_x spanning both the visible and surface wave range is plotted.

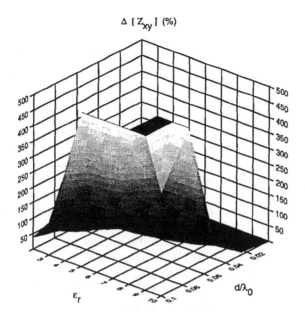

Figure 3.9: Plot of the average error between the polynomial approximation and the exact impedance, Z_{xy}, for a range of ϵ_r and d, when $\mu_r = 1.0$.

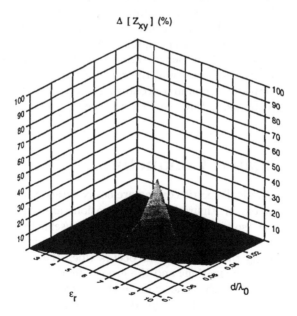

Figure 3.10: Plot of the average error between the ratio of polynomial approximation and the exact impedance, Z_{xy}, for a range of ϵ_r and d, when $\mu_r = 1.0$.

In this case the match points have been chosen to span the entire range of visible and surface wave space for k_x by choosing $k_x^{(1)} = 0.5\sqrt{\epsilon_r}k_0$ and $k_x^{(2)} = \sqrt{\epsilon_r}k_0$ for the ratio of polynomial approximation. For the single match point the value $k_x = 0.5\sqrt{\epsilon_r}k_0$ has been chosen. These plots illustrate the superiority of the ratio of polynomial approximation over the simple polynomial approximation and the SIBC. For $\sqrt{\epsilon_r\mu_r}d > \lambda_0/4$ the impedance becomes infinite for some value of k_x, indicating the presence of a real pole in k_x space. The ratio of polynomial approximation is capable of modeling this behavior whereas the simple polynomial approximation is not. The errors in the ratio of polynomial approximation do, however, tend to increase rapidly as the coating depth approaches one quarter wavelength in the dielectric.

Upon defining an acceptable average error in the impedance approximation, a range of validity for each of the approximate boundary conditions may be identified. In general, for lossless coatings, the errors in the approximate boundary condition tend to increase substantially as the coating approaches one quarter wavelength in depth. This range should be considered a maximum since the ability of the HOIBC to model the infinite planar problem accurately is only the first of two requirements that must be met in order for the boundary condition to be applicable to an arbitrary geometry. A second condition is that the locally planar approximation be valid at each point on the scattering body. The effects of curvature on this condition will be investigated for coated two-dimensional circular cylinders in the next chapter, as well as in the applications section of the book pertaining to arbitrary two-dimensional and three-dimensional geometries. The effects of discontinuities in the parameters of the coating on the locally planar approximation will be discussed in Chapter 5, where scattering by finite length grooves in an infinite ground plane is considered. Due to the superiority of the ratio of polynomial approximation, this higher order impedance boundary condition, rather than that based on the simple polynomial approximation, will be employed throughout the applications portion of this book.

3.2   Corrugated Conductors

The results for a simple dielectric coating are easily extended to the corrugated ground plane depicted in Figure 3.11. Corrugations are one method of simulating "hard" and "soft" surfaces [17], which are useful in a wide range of applications. The surface is said to be soft when the Poynting vector is zero for propagation along the surface. This occurs when the field is propagating perpendicular to the corrugations and the groove depth is adjusted to present an infinite impedance. On the other hand if the propagation direction is along the corrugations and the depth is adjusted appropriately a hard surface can be obtained. In this case fields of any polarization in the transverse plane can propagate along the surface in the direction of the corrugations. The standard model applied to a corrugated surface assumes a standing TEM wave in the

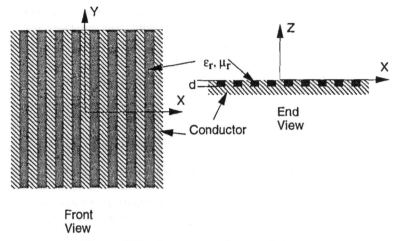

Figure 3.11: A corrugated ground plane.

corrugations and that the electric field parallel to the corrugations, E_y, is zero, resulting in,

$$Z_{yy}(k_x, k_y) = Z_{yx}(k_x, k_y) = 0, \qquad (3.22)$$

$$Z_{xy}(k_x, k_y) = -j\sqrt{\frac{\mu}{\epsilon}}\tan(k_z d), \qquad (3.23)$$

and

$$k_z = \sqrt{k^2 - k_y^2}. \qquad (3.24)$$

All of the remaining symbols have the same meanings as in the previous section. In addition, the final impedance term may be derived using reciprocity, as discussed in Appendix A, where it is shown that $Z_{xx} = -Z_{yy}$, and thus in this case

$$Z_{xx}(k_x, k_y) = 0. \qquad (3.25)$$

A simple model for the corrugations is to ignore any propagation along the groove, i.e., assume $k_y = 0$, and use a constant impedance tensor to model the behavior of the surface. This is an example of a Tensor Impedance Boundary Condition, TIBC. It has been used with great success for many years to model the behavior of corrugated horns and waveguides where the condition $k_y = 0$ is met exactly.

The above equations are the more general form of the boundary condition, valid when $k_y \neq 0$, and the corrugations are simulating either a hard or a soft surface. The above expression for $Z_{xy}(k_x, k_y)$ can then be expanded according to any of the methods described in Chapter 2, and then transformed into the spatial domain. For example, we may expand Z_{xy} as a ratio of polynomials,

$$Z_{xy}(k_y) \approx \frac{c_0 + c_1 k_y^2}{1 + c_2 k_y^2}. \qquad (3.26)$$

The methods discussed earlier may be applied to find the coefficients. Upon transforming into the spatial domain, the following two equations are found to describe the higher order boundary condition on a corrugated conductor,

$$\left(1 - c_2\frac{\partial^2}{\partial y^2}\right) E_x(x,y) = \left(c_0 - c_2\frac{\partial^2}{\partial y^2}\right) H_y(x,y) \tag{3.27}$$

and

$$E_y(x,y) = 0. \tag{3.28}$$

This boundary condition may be applied locally to each point on an arbitrarily shaped corrugated surface and may be used in conjunction with integral equations and the method of moments to investigate radiation and scattering problems involving corrugations acting as both hard and soft surfaces.

3.3   Planar Chiral Coatings

There has been a great deal of interest in the topic of chiral materials in recent years [28]. Chiral materials have the interesting property that right-handed and left-handed circularly polarized waves travel at different phase velocities in the material. This property gives rise to a number of interesting phenomena, which can be used to construct new devices [32]. In addition it has been shown that chiral coatings may be used to reduce the RCS of conducting bodies [33, 34]. Although the dyadic Green's function for chiral media is available [35], it is considerably more complicated than the free space Green's function. As was discussed in the introduction it would be possible to compute the scattering from chiral coated bodies without the use of this Green's function if a suitable higher-order boundary condition could be found for the chiral coating. Furthermore, the number of unknowns may be reduced relative to that required in an exact formulation of the scattering problem. In addition, the boundary condition can be used in high frequency diffraction studies. Application of the Generalized Impedance Boundary Condition to a slab of chiral material with no conductor backing has been discussed briefly in Ref. [36]. In the following sections the appropriate higher order boundary condition for a planar chiral coating will be derived using the general method outlined earlier. It will be demonstrated that consideration of this complicated coating is only slightly more complicated than the dielectric coating when the spectral domain approach is employed. The accuracy of the resulting HOIBC will then be demonstrated for several example coatings.

3.3.1   Exact Spectral Domain Boundary Conditions

Assuming $e^{j\omega t}$ time dependence, a chiral medium is characterized by Maxwell's equations

$$\nabla \times \mathbf{E} = -j\omega\mathbf{B}, \ \nabla \times \mathbf{H} = j\omega\mathbf{D}, \qquad (3.29)$$

and the constitutive relations

$$\mathbf{D} = \epsilon\mathbf{E} - j\gamma_c\mathbf{B}, \ \mathbf{H} = \frac{1}{\mu}\mathbf{B} - j\gamma_c\mathbf{E}. \qquad (3.30)$$

The chiral admittance, γ_c, is responsible for the unique properties of the medium, which in the special case $\gamma_c = 0$ reduces to a simple dielectric.

The first step in determining the appropriate higher order boundary condition for the conductor-backed layer of chiral material shown in Figure 3.12 is to

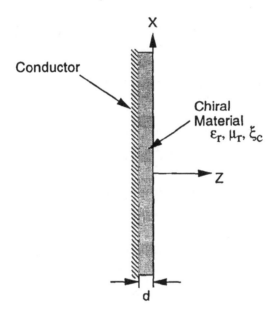

Figure 3.12: A planar chiral layer.

evaluate the exact impedance tensor. Once again the impedances for the case $k_y = 0$ will be derived, since these along with the rotational invariance of the coating are satisfactory for deriving the higher order boundary conditions. To proceed we require the four wavenumbers, electric field vectors, and magnetic field vectors of Eqs. (2.1) and (2.2). For the chiral media the appropriate four waves are forward $(+z)$ and reverse $(-z)$ propagating right-hand (RCP) and left-hand circularly polarized (RCP) plane waves. The chiral medium is distinguished from a dielectric medium by the fact that the RCP and LCP waves have different wavenumbers, which are given in Ref. [35],

$$\left.\begin{array}{c} k_r \\ k_l \end{array}\right\} = \omega\sqrt{\mu\epsilon_c} \pm \omega\mu\gamma_c, \qquad (3.31)$$

with

$$\epsilon_c = \epsilon + \mu\gamma_c^2. \tag{3.32}$$

The appropriate tangential field components for the case $k_y = 0$ are found to be

$$\mathbf{e}_1 = j\frac{k_z^{(1)}}{k_r}\hat{\mathbf{x}} + \hat{\mathbf{y}}, \quad \mathbf{h}_1 = \frac{j}{\eta_c}\mathbf{e}_1, \quad k_z^{(1)} = \sqrt{k_r^2 - k_x^2}, \tag{3.33}$$

$$\mathbf{e}_2 = j\frac{k_z^{(2)}}{k_r}\hat{\mathbf{x}} + \hat{\mathbf{y}}, \quad \mathbf{h}_2 = \frac{j}{\eta_c}\mathbf{e}_2, \quad k_z^{(2)} = -\sqrt{k_r^2 - k_x^2}, \tag{3.34}$$

$$\mathbf{e}_3 = -j\frac{k_z^{(3)}}{k_l}\hat{\mathbf{x}} + \hat{\mathbf{y}}, \quad \mathbf{h}_3 = -\frac{j}{\eta_c}\mathbf{e}_1, \quad k_z^{(3)} = \sqrt{k_l^2 - k_x^2}, \tag{3.35}$$

and

$$\mathbf{e}_4 = -j\frac{k_z^{(4)}}{k_l}\hat{\mathbf{x}} + \hat{\mathbf{y}}, \quad \mathbf{h}_4 = -\frac{j}{\eta_c}\mathbf{e}_1, \quad k_z^{(4)} = -\sqrt{k_l^2 - k_x^2}, \tag{3.36}$$

where $\eta_c = \sqrt{\mu/\epsilon_c}$.

Upon substituting these expressions into the general expression for the exact impedance tensor, Eq. (2.12), and performing the indicated manipulations the following exact impedances may be determined

$$Z_{xx}(k_x, 0) = \frac{j\eta_c k_x^2(k_r^2 - k_l^2)\sin k_{rz}d\sin k_{lz}d}{D(k_x, 0)}, \tag{3.37}$$

$$Z_{xy}(k_x, 0) = \frac{-2j\eta_c(k_{lz}^2 k_{rz}k_r\cos k_{rz}d\sin k_{lz}d + k_{rz}^2 k_{lz}k_l\cos k_{lz}d\sin k_{rz}d)}{D(k_x, 0)}, \tag{3.38}$$

$$Z_{yx}(k_x, 0) = \frac{2j\eta_c k_r k_l(k_{rz}k_l\cos k_{rz}d\sin k_{lz}d + k_{lz}k_r\cos k_{lz}d\sin k_{rz}d)}{D(k_x, 0)}, \tag{3.39}$$

with

$$D(k_x, 0) = 2k_r k_{rz}k_l k_{lz}(1 + \cos k_{rz}d\cos k_{lz}d) - (k_{zl}^2 k_r^2 + k_{zr}^2 k_l^2)\sin k_{rz}d\sin k_{lz}d, \tag{3.40}$$

and

$$Z_{yy}(k_x, 0) = -Z_{xx}(k_x, 0), \tag{3.41}$$

as is required by reciprocity.

Several points about these exact impedance functions are worth noting. If the chiral admittance γ_c becomes zero, then $k_r = k_l$, the impedances Z_{xx} and Z_{yy} vanish, and the other two impedances reduce to those derived in the previous section for a dielectric layer. As will be shown in the examples, the presence of the nonzero Z_{xx} and Z_{yy} produces a cross polarized reflection from the chiral layer at oblique angles of incidence.

3.3.2 Higher Order Impedance Boundary Conditions

The coefficients are then determined using the general procedure described in Chapter 2. In this case we have $Z_{xx}(0,0) = Z_{yy}(0,0) = c_0^5 = 0$, and the remainder of the coefficients are in general nonzero. The differential equations have exactly the form shown in Eqs. (2.28) and (2.29). It remains to be shown that these differential equations are a good model for the behavior of the coating. This is the subject of the next section.

3.3.3 Examples

As a first example a chiral coating with parameters $\epsilon_r = 4.0$, $\mu_r = 1.0$, $\gamma_c = 0.003\Omega^{-1}$, and depth $d = 0.1\lambda_0$ is considered. Figure 3.13 shows the exact

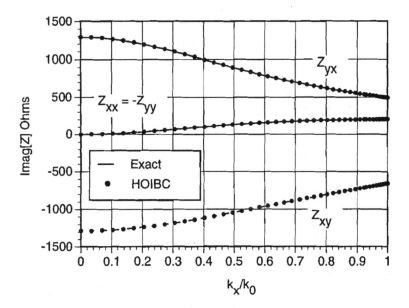

Figure 3.13: Impedances for a planar chiral layer with $\epsilon_r = 4.0$, $\mu_r = 1.0$, $\gamma_c = 0.003\Omega^{-1}$, and $d = 0.1\lambda_0$.

impedance terms, as well as the approximate impedances computed using the HOIBC. Although the impedances have been forced to be equal only at $\theta = 0$ degrees ($k_x = 0$), 90 degrees ($k_x = k_0$) and $\theta = 30$ degrees ($k_x = k_0/2$) the impedance approximations are excellent for all angles of incidence in the visible region and are also excellent approximations for all k_x resulting in z propagation in the layer as well. In this particular case the complicated behavior of the coating is captured quite well by the second order differential equations.

As was discussed earlier the existence of the impedances Z_{xx} and Z_{yy} can be attributed to γ_c. Since these impedances are zero for normal incidence $k_x = 0$, it is impossible to determine if there is any chirality present in the coating from examining the reflection of a normally incident plane wave. Thus, approximate boundary conditions that rely on a normally incident plane wave approximation such as the SIBC or TIBC are incapable of simulating the behavior of the chiral coating. The magnitudes of the reflection coefficients are plotted in Figure 3.14.

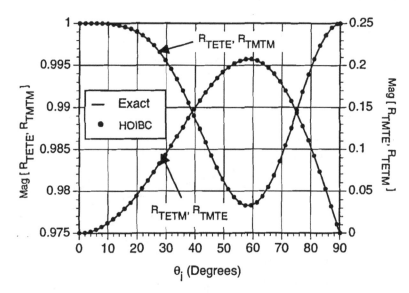

Figure 3.14: Magnitude of the reflection coefficients for a planar chiral layer with $\epsilon_r = 4.0$, $\mu_r = 1.0$, $\gamma_c = 0.003\Omega^{-1}$, and $d = 0.1\lambda_0$.

The reflection coefficient calculations depend directly on the accuracy of the impedance approximations of Figure 3.13. Therefore, given the excellent results of Figure 3.13, the accuracy of the HOIBC demonstrated in Figure 3.14 is to be expected. Errors of much less than 1% are seen over the entire range of incident angles, whereas the TIBC (SIBC) would fail to produce any cross polarized reflection, since $Z_{xx} = Z_{yy} = 0$ at normal incidence. As is noted on the figure, reciprocity dictates that $R_{TETM} = R_{TMTE}$, and conservation of power (lossless layer) then gives $|R_{TETE}| = |R_{TMTM}|$.

The second example is for a thin chiral coating with parameters $\epsilon_r = 1.0$, $\mu_r = 2.0$, and $\gamma_c = 0.005\Omega^{-1}$, and depth $d = 0.02\lambda_0$. The magnitude and phase of the reflection coefficients are plotted in Figure 3.15 and Figure 3.16 respectively. As with all planar chiral coatings the cross-polarized reflection is zero for normal and grazing incidence, reaching a maximum at an angle determined by the parameters of the coating. Due to the larger values of γ_c and μ_r the RCP and LCP propagation coefficients for waves in this chiral material

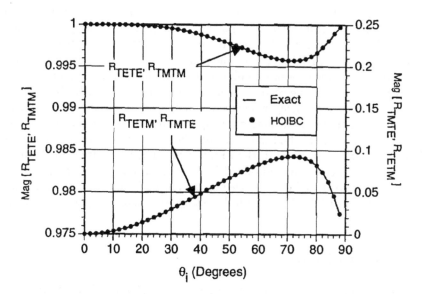

Figure 3.15: Magnitude of the reflection coefficients for a chiral layer with $\epsilon_r = 1.0$, $\mu_r = 2.0$, $\gamma_c = 0.005\Omega^{-1}$, and $d = 0.02\lambda_0$.

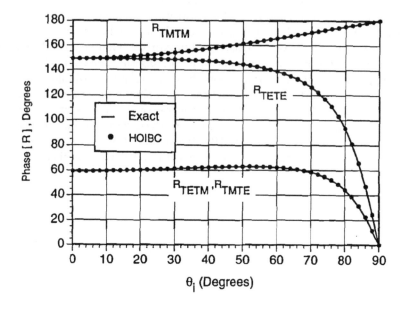

Figure 3.16: Phase of the reflection coefficients for a chiral layer with $\epsilon_r = 1.0$, $\mu_r = 2.0$, $\gamma_c = 0.005\Omega^{-1}$, and $d = 0.02\lambda_0$.

are more distinct than those of the previous example. Thus, this thin coating still generates a substantial cross-polarized component for angles of incidence near 70 degrees. Figure 3.15 shows the excellent agreement between the exact solution and the HOIBC solution. The phase of the reflection coefficients is also in excellent agreement, well within a degree, as illustrated in Figure 3.16.

As for the case of dielectric coatings, the accuracy of the HOIBC generally degrades with increasing layer thickness and index of refraction. For thick layers the impedance terms are no longer well approximated by a simple ratio of second order polynomials, and higher order boundary conditions are required to simulate the coating. In general, the results presented in Figure 3.13-Figure 3.16 are indicative of the results obtained for a large number of coating parameters, and the spectral domain approach has proven to be an excellent method for determining the HOIBC for a chiral coating less than one quarter wavelength in depth.

3.4 Conclusions

In this chapter the general theory presented in Chapter 2 has been used to derive higher order impedance boundary conditions for planar dielectric and chiral coatings as well as corrugated conductors. For dielectric coatings the improvement of the HOIBC relative to the SIBC was demonstrated for several examples. The implications of the HOIBC for the surface wave behavior of the coating were discussed in a straightforward manner. The HOIBC were shown to be capable of modeling the complicated behavior of a chiral layer quite accurately, whereas the existing approximate boundary conditions such as the SIBC and TIBC are incapable of predicting this behavior. Construction of the appropriate HOIBC for this complicated coating was obtained with very little additional effort relative to the case of a simple dielectric coating. The HOIBC model for the corrugated conductor was shown to be a simple extension to that for a dielectric coating. In contrast to the existing boundary conditions for this surface treatment, the HOIBC is capable of automatically predicting the correct behavior of the corrugations when it is behaving as either a hard or soft surface.

Chapter 4

Boundary Conditions for Curved Dielectric and Chiral Coatings

In this chapter the applicability of the higher order impedance boundary conditions to the solution of scattering by coated conducting circular cylinders will be addressed. Coatings made up of simple dielectric as well as chiral material will be considered. The examples will illustrate the effects of curvature on the accuracy of the higher order impedance boundary conditions that are based on a planar canonical problem. Examination of the exact series solution for this problem leads directly to a means of correcting the planar higher order impedance boundary conditions for the effects of curvature.

4.1 Scattering by Dielectric Coated Circular Cylinders using Planar HOIBC

Consider the problem of scattering by the coated conducting circular cylinder depicted in Figure 4.1. The radius of the inner conductor is a, and the thickness of the coating is $b-a$. It is assumed that the incident field is propagating normal to the cylinder, $\partial/\partial z = 0$, and is polarized either with the electric field in the z

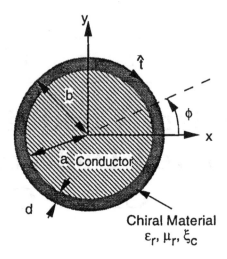

Figure 4.1: A chiral coated conducting circular cylinder.

direction (E Polarization or TM_z), or with the magnetic field in the z direction (H Polarization or TE_z). In order to illustrate several key points the case of a simple dielectric coating will be considered, and later it will be generalized to include the effects of finite chirality.

An exact solution of the scattering problem depicted in Figure 4.1 is obtained by expanding the incident field, the scattered field outside the cylinder, and the total field inside the cylinder coating in terms of a series of cylindrical wave functions and applying the appropriate boundary conditions at each interface [37]. The exact series solution as well as those associated with the HOIBC and SIBC formulations of the problem are described using a common framework in Appendix E.

The pertinent geometry for the higher order boundary condition solution to the scattering problem is depicted in Figure 4.2. Note that as illustrated in the figure the z axis has been chosen to be normal to the cylinder at the point of interest. This allows the application of the planar boundary conditions derived in the previous chapter without a coordinate rotation. At each point on the cylinder the tangential field components are postulated to satisfy the higher order impedance boundary condition derived previously for the coated ground plane with

$$\frac{\partial}{\partial y} \to 0 \tag{4.1}$$

and

$$\frac{\partial}{\partial x} \to -\frac{\partial}{b\partial \phi}, \tag{4.2}$$

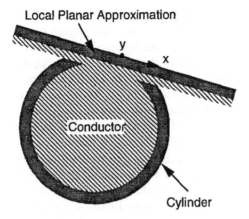

Figure 4.2: Geometry for application of the planar higher order impedance boundary condition on a coated circular cylinder.

$$E_x \rightarrow -E_\phi, H_x \rightarrow -H_\phi, \tag{4.3}$$

$$E_y \rightarrow E_z, \text{and } H_y \rightarrow H_z, \tag{4.4}$$

where it should be noted that since the positive t direction has been chosen to correspond to the positive x direction in the planar case, $\hat{t} = -\hat{\phi}$. Substituting into Eqs. (3.19) and (3.20) then gives

$$\left(1 - c_4^1 \frac{\partial^2}{b^2 \partial \phi^2}\right) E_z(\phi) = \left(c_0^6 - c_4^6 \frac{\partial^2}{b^2 \partial \phi^2}\right) H_\phi(\phi) \tag{4.5}$$

and

$$\left(1 - c_3^1 \frac{\partial^2}{b^2 \partial \phi^2}\right) E_\phi(\phi) = \left(-c_0^6 + c_3^6 \frac{\partial^2}{b^2 \partial \phi^2}\right) H_z(\phi). \tag{4.6}$$

Furthermore, since the incident and scattered fields are described in terms of a Fourier series in ϕ, see Appendix E, $\partial/\partial\phi = jn$ for the component that has a ϕ variation of $e^{jn\phi}$.

Substituting into the previous expressions we find that the tangential field components are related to each other through impedances that depend on the Fourier component number, n,

$$E_z = -\frac{c_0^6 + \left(c_4^6 n^2 / b^2\right)}{1 + \left(c_4^1 n^2 / b^2\right)} H_\phi \tag{4.7}$$

and

$$E_\phi = \frac{c_0^6 + \left(c_3^6 n^2 / b^2\right)}{1 + \left(c_3^1 n^2 / b^2\right)} H_z. \tag{4.8}$$

Note that when the Standard Impedance Boundary Condition is employed, only the coefficient c_0^6 is non-zero, and the impedance becomes independent of Fourier mode number. This illustrates the key feature of the higher order impedance boundary conditions that makes them superior to the SIBC. Since derivatives of the fields are included in the boundary condition, the HOIBC can detect any propagation along the dielectric to free space boundary and correct the impedances for normal incidence appropriately. The SIBC on the other hand does not have the capability to detect this propagation, employing an impedance that is accurate only when such propagation is negligible.

Since the coefficients appearing in the HOIBC were derived by considering the planar canonical problem it is expected that the solution should be most accurate for cylinders with large radii of curvature and thin coatings, where the geometrical approximation depicted in Figure 4.2 is a good one.

In order to illustrate these points scattering by three typical coated cylinders will be considered next. The details of the series solution for scattering by a circular cylinder with an arbitrary boundary condition is outlined in Appendix E. Figure 4.3 shows the bistatic radar cross section for a coated conducting cylinder with inner radius λ_0, coating thickness $0.075\lambda_0$, and coating parameters $\epsilon_r = 6.0$ and $\mu_r = 1.0$, and H polarization. The exact series solution is presented along with the HOIBC and SIBC solutions. Nearly all of the details

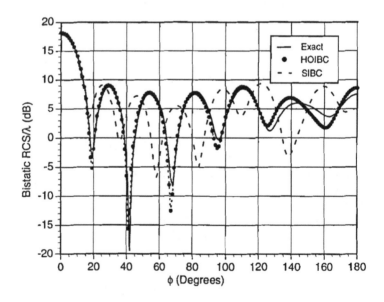

Figure 4.3: Bistatic radar cross section for a coated conducting cylinder with $a=\lambda_0$, $d=0.075\lambda_0$, $\epsilon_r = 6.0$, $\mu_r = 1.0$, and H polarization. Backscatter direction is 180 degrees.

of the scattered field are computed correctly by the HOIBC, while the SIBC results are quite inaccurate. The slight inaccuracy in the HOIBC solution can be attributed to the "locally planar" approximation. This is illustrated by the results of Figure 4.4, where the radius of the inner conductor is increased to $3.0\lambda_0$. As expected, the accuracy of the HOIBC solution is improved

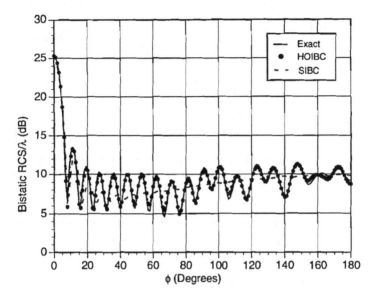

Figure 4.4: Bistatic radar cross section for a coated conducting cylinder with $a=3.0\lambda_0$, $d=0.075\lambda_0$, $\epsilon_r = 6.0$, $\mu_r = 1.0$, and H polarization. Backscatter direction is 180 degrees.

while the SIBC solution is still inadequate. As was mentioned previously, the SIBC is known to be a good approximation for highly lossy coatings. This is illustrated in Figure 4.5 where the same cylinder geometry of Figure 4.3 is used, but the coating is made lossy, $\epsilon_r = 6.0 - j6.0$. In this case the HOIBC and SIBC solutions converge to the same result, which is in excellent agreement with the exact solution. For this set of coating parameters the impedance is essentially independent of tangential wavenumber (Fourier component), and only c_0^6 is found to be significant when the HOIBC coefficients are determined. Thus, the HOIBC solution reverts to that of the SIBC under the appropriate circumstances.

The next several figures are included to illustrate the relative accuracy of the HOIBC and SIBC for a range of coating parameters given a fixed cylinder size. Figure 4.6 plots the average error in the scattered field in dB relative to the exact backscattered field as a function of ϵ_r and d when the cylinder inner radius is λ_0 and $\mu_r = 1.0$. Results for H polarization are plotted. Several conclusions may be drawn from these results. In all cases the results of the HOIBC are

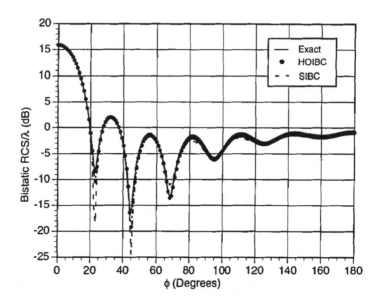

Figure 4.5: Bistatic radar cross section for a coated conducting cylinder $a=\lambda_0$, $d=0.075\lambda_0$, $\epsilon_r = 6.0 - j6.0$, $\mu_r = 1.0$, and H polarization. Backscatter direction is 180 degrees.

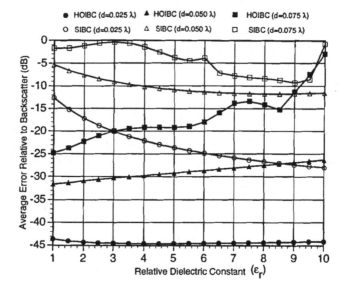

Figure 4.6: Average error in the scattered field (dB relative to the exact backscattered field) for H polarization as a function of ϵ_r and d when the cylinder inner radius is λ_0 and $\mu_r = 1.0$.

superior to those of the SIBC. For a given coating depth the HOIBC error first decreases as the dielectric constant increases, and then increases again as the coating approaches a depth of approximately one quarter wavelength in the dielectric. On the other hand, the SIBC accuracy improves monotonically for increasing dielectric constant given a fixed coating thickness.   Results for a

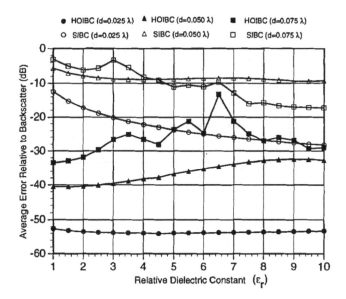

Figure 4.7: Average error in the scattered field (dB relative to the exact back-scattered field) for H polarization as a function of ϵ_r and d when the cylinder inner radius is $3.0\lambda_0$ and $\mu_r = 1.0$.

larger cylinder, $a = 3.0\lambda_0$, are presented in Figure 4.7. As expected the error of the HOIBC is improved relative to the $a = \lambda_0$ case and is superior to the SIBC in all cases. For $d = 0.075\lambda_0$ a small region of higher average error is found near $\epsilon_r = 6.5$. This is due to significant excitation of a high order Fourier component that is not predicted accurately by the HOIBC. This effect will be discussed in detail when magnetic coatings are considered, since the effect is even more pronounced when $\mu_r \neq 1$.

Results for E polarization are presented in Figure 4.8. As can be noted from the plot both the SIBC and HOIBC solutions are quite acceptable over the range of coating parameters plotted, and the results for E polarization are superior to those for H polarization in all cases. This is due to the lack of surface wave excitation by this particular polarization of the field and a corresponding lack of structure in the scattered field pattern.

The accuracy of the HOIBC and SIBC for the case of a one wavelength cylinder with a magnetic coating is illustrated in Figure 4.9. As can be seen from the figure these results are quite different from those of a coating with $\mu_r = 1$. The

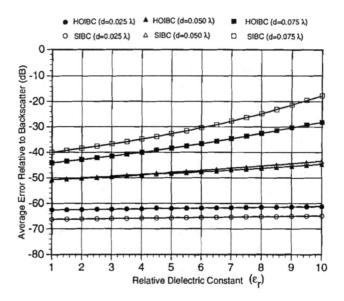

Figure 4.8: Average error in the scattered field (dB relative to the exact back-scattered field) for E polarization as a function of ϵ_r and d when the cylinder inner radius is $3.0\lambda_0$ and $\mu_r = 1.0$.

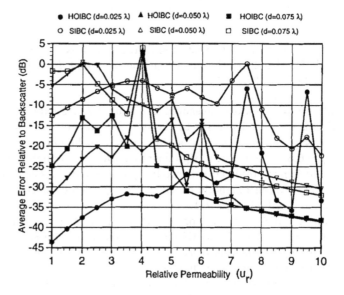

Figure 4.9: Average error in the scattered field (dB relative to the exact back-scattered field) for H polarization as a function of μ_r and d when the cylinder inner radius is λ_0 and $\epsilon_r = 1.0$.

performance of the HOIBC is still superior to that of the SIBC, but the region of maximum error is shifted to lower coating depths relative to the $\epsilon_r \neq 1$ case. Unlike the ϵ_r case substantial errors occur in the scattered field even for modest coating depths. This is illustrated in Figure 4.10 where the bistatic RCS for the specific case $\epsilon_r = 1$, $\mu_r = 3$, $d = 0.02\lambda_0$, and $a = \lambda_0$. As Figure 4.9 indicates the scattered field predicted by the SIBC has substantial error, even for this thin coating. The spikes in Figure 4.9 indicate that for particular choices of coating

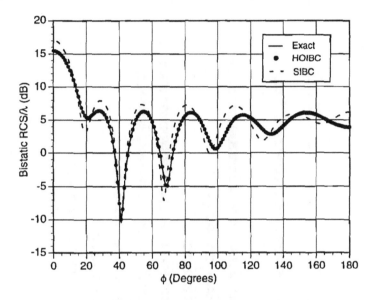

Figure 4.10: Bistatic radar cross section for a coated conducting cylinder with $a = \lambda_0$, $d = 0.02\lambda_0$, $\epsilon_r = 1.0$, $\mu_r = 3.0$, and H polarization. Backscatter direction is 180 degrees.

parameters the planar HOIBC becomes highly inaccurate. These regions of high error are due to the fact that for these particular choices of coating parameters a higher order term (surface wave) in the series solution becomes resonantly excited. In order to correctly predict this resonant excitation the impedance must be accurately represented for the particular Fourier component involved. The planar HOIBC is not an accurate enough model to correctly predict this behavior. This is illustrated in Figure 4.11 where the coefficients of the Fourier series representation of the reflected wave computed using an exact solution as well as the error between the planar HOIBC terms and the exact terms are presented for the case $a=\lambda_0$, $d=0.25\lambda_0$, $\mu_r = 7.5$, and $\epsilon_r = 1.0$. As can be seen in the figure in this case the $n=10$ term in the series is highly excited but the planar HOIBC solution fails to predict it correctly, although the HOIBC is fairly accurate in the magnitudes of all of the other terms in the series.

This parametric study can be used to make some predictions about the ap-

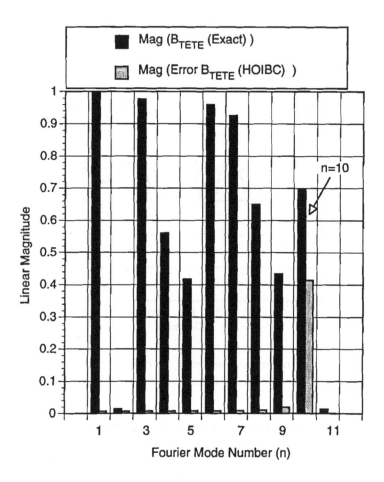

Figure 4.11: Exact terms in the Fourier series expansion of the reflected field and error between the planar HOIBC terms and the exact terms for a coated circular cylinder when $a=\lambda_0$, $d=0.25\lambda_0$, $\mu_r = 7.5$, and $\epsilon_r = 1.0$ and the incident field is H polarized.

plication of the HOIBC based on the planar canonical problem to scattering by bodies of more general shape. For dielectric coatings the error in the scattered field computed using the HOIBC is expected to be maximum for coatings approaching one quarter of a wavelength deep. For magnetic coatings the depth corresponding to the maximum error should be somewhat less. The scattered field computed using the HOIBC is capable of predicting the details of the sidelobe structure of the scattered field with reasonable accuracy even when the effects of curvature are ignored, whereas the SIBC is inadequate in this respect. As expected, the accuracy of the planar HOIBC increases as the radii of curvature on the scattering body increase and the coating depth decreases. Additional investigations have shown that the accuracy of the HOIBC improves for lossy dielectric and magnetic coatings.

The parametric study of scattering by dielectric coated cylinders clearly illustrates one of the major limitations of the HOIBC as presented up to this point, the locally planar approximation. The next section will remove this approximation by replacing the planar canonical problem by that of the coated circular cylinder. In this manner HOIBC that include the effects of curvature in one dimension may be determined for dielectric and chiral coatings.

4.2   Higher Order Impedance Boundary Conditions for Curved Coatings

The spectral domain approach for finding the HOIBC may be extended to include the effects of curvature in one dimension quite systematically. The next few sections will outline the procedure for including the effects of curvature, and will present some examples of scattering by circular cylinders with dielectric as well as chiral coatings, demonstrating the increased accuracy of the HOIBC that includes curvature relative to the planar HOIBC.

4.2.1   Exact Spectral Domain Boundary Conditions

The methods for determining the HOIBC discussed earlier in the chapter may be extended to include curvature in one dimension by substituting the canonical problem of scattering at a chiral coated circular cylinder, Figure 4.1, for the planar problem. By simplifying setting the chirality of the layer to zero the results for a simple curved dielectric layer are recovered.

Consider the field inside the layer of material of thickness $d = b - a$. For a fixed longitudinal wavenumber k_z the field inside the coating is written as a Fourier series in ϕ with each term consisting of a sum of four waves with unknown magnitudes,

$$\mathbf{E}\left(k_z, r\right) = \sum_{i=1}^{4} \sum_{n=-\infty}^{\infty} c_{in} \mathbf{e}_i\left(n, r\right) e^{-jk_z z} e^{jn\phi}, \tag{4.9}$$

and

$$\mathbf{H}(k_z, r) = \sum_{i=1}^{4} \sum_{n=-\infty}^{\infty} c_{in} \mathbf{h}_i(n, r) e^{-jk_z z} e^{jn\phi}. \tag{4.10}$$

The development will be carried out assuming that the coating has finite chirality, the dielectric coating case being a special case when the chirality vanishes. As before, the above expansion is valid regardless of the angle of incidence of the incident field or its polarization. The eight vector fields $\mathbf{e}_i(n, r)$ and $\mathbf{h}_i(n, r)$ are the characteristic cylindrical waves in the medium, and in this case they involve Hankel functions rather than the trigonometric functions that appear in the planar case. Due to the orthogonality of the $e^{jn\phi}$ terms, an exact impedance boundary condition involving the four tangential field components may be obtained for each value of n, which may be written in terms of an impedance tensor as

$$\left[\begin{array}{c} \tilde{E}_\phi(n, k_z) \\ \tilde{E}_z(n, k_z) \end{array} \right] = \left[\begin{array}{cc} Z_{\phi\phi}(n, k_z) & Z_{\phi z}(n, k_z) \\ Z_{z\phi}(n, k_z) & Z_{zz}(n, k_z) \end{array} \right] \left[\begin{array}{c} \tilde{H}_\phi(n, k_z) \\ \tilde{H}_z(n, k_z) \end{array} \right]. \tag{4.11}$$

The development of this impedance tensor parallels that of the planar case, where the z and ϕ components of the field are of interest rather than the x and y components. For the present study we confine ourselves to the case of normal incidence, i.e., $k_z = 0$. In this case the impedance terms for a chiral coated cylinder may be computed using the field expansions detailed in [38]. This method can be used to determine the input impedance for multilayer chiral coatings on circular cylinders using simple transmission line techniques.

Repeating the procedure detailed in Chapter 2, but considering the tangential fields in the cylindrical layer, we arrive at the following solution for the impedance tensor,

$$[Z] = -j\eta_c \left([M_{12}(b)] + [M_{34}(b)] [M_{34}(a)]^{-1} [M_{12}(a)] \right)$$
$$\left([M_{12}(b)] - [M_{34}(b)] [M_{34}(a)]^{-1} [M_{12}(a)] \right)^{-1}, \tag{4.12}$$

with

$$[M_{12}(r)] = \left[\begin{array}{cc} -J_n'(k_r r) & -H_n^{(2)\prime}(k_r r) \\ J_n(k_r r) & H_n^{(2)}(k_r r) \end{array} \right] \tag{4.13}$$

and

$$[M_{34}(r)] = \left[\begin{array}{cc} J_n'(k_l r) & H_n^{(2)\prime}(k_l r) \\ J_n(k_l r) & H_n^{(2)}(k_l r) \end{array} \right]. \tag{4.14}$$

Here J_n is the Bessel function of the first kind, $H_n^{(2)}$ is the Hankel function of the second kind, and the prime indicates a derivative with respect to the argument. The wavenumbers k_r and k_l are the right-handed and left-handed wavenumbers for the chiral media and are described in Chapter 3. The impedance terms in Eq. (4.12) are now complicated expressions involving Hankel and Bessel functions and the coating parameters a, b, ϵ_r, μ_r, and γ_c. The algebraic expressions for these impedances are omitted for the sake of brevity. When $\gamma_c = 0$ they reduce to the impedances for a simple dielectric layer,

$$Z_{\phi z}(n) = j\eta \frac{H_n^{(2)'}(ka)H_n^{(1)'}(kb) - H_n^{(1)'}(ka)H_n^{(2)'}(kb)}{H_n^{(2)'}(ka)H_n^{(1)}(kb) - H_n^{(1)'}(ka)H_n^{(2)}(kb)}, \qquad (4.15)$$

$$Z_{z\phi}(n) = -j\eta \frac{H_n^{(1)}(ka)H_n^{(2)}(kb) - H_n^{(2)}(ka)H_n^{(1)}(kb)}{H_n^{(1)}(ka)H_n^{(2)'}(kb) - H_n^{(2)}(ka)H_n^{(1)'}(kb)}, \qquad (4.16)$$

and

$$Z_{\phi z}(n) = Z_{z\phi}(n) = 0, \qquad (4.17)$$

for $k_z = 0$. It should be noted that the above expressions may be used to correct the SIBC for the effects of curvature. Evaluating the appropriate impedance for the case of zero tangential wavenumber, $n = 0$, and using it as the SIBC impedance will provide a curvature correction to the planar SIBC.

We note that the wavenumber k_z is analogous to k_y for the planar case. We may define an azimuthal wavenumber $k_t = -n/b$, which corresponds to propagation in the tangential direction, as shown in Figure 4.1. This wavenumber is analogous to the wavenumber k_x in the planar case. The exact boundary condition for the case $k_z = 0$ may be written entirely in terms of the tangential wavenumber k_t as

$$\begin{bmatrix} \tilde{E}_t(k_t b, 0) \\ \tilde{E}_z(k_t b, 0) \end{bmatrix} = \begin{bmatrix} Z_{tt}(k_t b, 0) & Z_{tz}(k_t b, 0) \\ Z_{zt}(k_t b, 0) & Z_{zz}(k_t b, 0) \end{bmatrix} \begin{bmatrix} \tilde{H}_t(k_t b, 0) \\ \tilde{H}_z(k_t b, 0) \end{bmatrix}. \qquad (4.18)$$

4.2.2 Higher Order Impedance Boundary Conditions

We may now proceed to approximate this boundary condition in terms of polynomials in k_t and k_z. In this case arbitrary rotational invariance does not apply, but 180 degree rotational invariance does apply, eliminating the linear k_z and k_t terms from the polynomials. Upon performing the inverse Fourier transform on the approximate boundary condition, $k_z \to j\partial/\partial z$ and $k_t \to j\partial/\partial t$, and a pair of differential equations relating the t and z components of the field and their t and z derivatives results. These are then the appropriate differential equations to apply on a three-dimensional scatterer at a location where one of the principal radii of curvature is infinite and the other is b.

In this case of normal incidence $k_z = 0$ and the polynomials have the following form

$$P_1(k_t) = 1 + c_3^1 k_t^2, \quad P_4(k_t) = 1 + c_3^4 k_t^2, \qquad (4.19)$$

$$P_2(k_t) = c_3^2 k_t^2, \; P_3(k_t) = c_3^3 k_t^2, \tag{4.20}$$

$$P_5(k_t) = c_0^5 + c_3^5 k_t^2, \; P_8(k_t) = c_0^8 + c_3^8 k_t^2, \tag{4.21}$$

and

$$P_6(k_t) = c_0^6 + c_3^6 k_t^2, \; P_7(k_t) = c_0^7 + c_3^7 k_t^2. \tag{4.22}$$

Again the twelve coefficients are determined by satisfying the cylindrical equivalent of Eq. (2.18) exactly at three values of k_t. Two additional equations result since for this case the lack of rotational invariance allows $Z_{tt} = -Z_{zz} \neq 0$, as well as $Z_{tz} \neq -Z_{zt}$ at $k_t = k_z = 0$. The coefficients c_0^n are determined immediately at $k_t = 0$, and the remaining eight are obtained by satisfying the cylindrical equivalent of Eq. (2.18) at two distinct values of k_t, typically $k_t = k_0/2$, and $k_t = k_0$.

Using the Fourier transform the polynomial approximation results in the following higher order boundary conditions in the spatial domain,

$$\left(1 - c_3^1 \frac{\partial^2}{\partial t^2}\right) E_t(t) - c_3^2 \frac{\partial^2 E_z(t)}{\partial t^2} = \left(c_0^5 - c_3^5 \frac{\partial^2}{\partial t^2}\right) H_t(t) + \left(c_0^6 - c_3^6 \frac{\partial^2}{\partial t^2}\right) H_z(t) \tag{4.23}$$

and

$$-c_3^3 \frac{\partial^2 E_t(t)}{\partial t^2} + \left(1 - c_3^4 \frac{\partial^2}{\partial t^2}\right) E_z(t) = \left(c_0^7 - c_3^7 \frac{\partial^2}{\partial t^2}\right) H_t(t) + \left(c_0^8 - c_3^8 \frac{\partial^2}{\partial t^2}\right) H_z(t). \tag{4.24}$$

These are the appropriate equations for curved chiral coatings. As discussed in Appendix E, these two equations are used to determine the coefficients of the scattered cylindrical waves when a plane wave is incident on a coated circular cylinder. As we will see the effects of curvature can be very significant for a chiral coating, particularly for the cross polarized reflection. This is due to the fact that for a curved chiral coating Z_{tt} and Z_{zz} are nonzero even for normal incidence $k_t = 0$, and therefore c_0^5 and c_0^8 are nonzero for the curved coating.

4.2.3   Examples

In order to demonstrate the increased accuracy of boundary conditions that include the effects of curvature we apply them to the dielectric coated cylinder shown in Figure 4.3. In Figure 4.12 is shown the bistatic RCS of a coated cylinder with a conductor radius of λ_0, coating thickness of $0.075 \, \lambda_0$, $\mu_r = 1.0$, and $\epsilon_r = 6.0$, for H polarization. The exact solution as well as approximate solutions based on the planar HOIBC and the HOIBC that includes curvature are plotted. While the planar results are quite acceptable, including the effects of curvature in the boundary condition produces essentially the exact solution. For this fairly large cylinder and a thin coating the effect of including the curvature is to modify the coefficients only slightly.

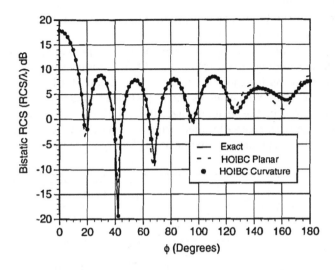

Figure 4.12: Bistatic radar cross section for a coated conducting cylinder with inner radius λ_0, coating thickness $0.075\lambda_0$, and coating parameters $\epsilon_r = 6.0$ and $\mu_r = 1.0$, and H polarization. Backscatter direction is 180 degrees.

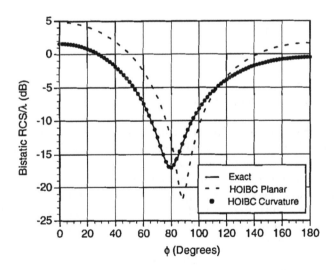

Figure 4.13: Bistatic radar cross section for a coated conducting cylinder with $a=0.1\lambda_0$, $d=0.075\lambda_0$, $\epsilon_r = 6.0$, $\mu_r = 1.0$, and H polarization. Backscatter direction is 180 degrees.

For smaller cylinders with larger curvature it is imperative to include curvature effects in the boundary condition. This illustrated in Figure 4.13 where the same coating as in the previous example is applied to a cylinder with a conductor radius of $0.1\lambda_0$. In this case the locally planar approximation is a poor one, but including curvature in the HOIBC produces essentially the exact solution. The reason for the inaccuracy of the planar HOIBC solution is illustrated in Figure 4.14, which shows the impedance terms versus the normalized wavenumber k_t/k_0 in the visible range, $0 \leq k_t \leq k_0$. The exact impedances are shown along with those obtained using the HOIBC that includes curvature effects. Clearly the planar approximation is highly inaccurate in predicting the

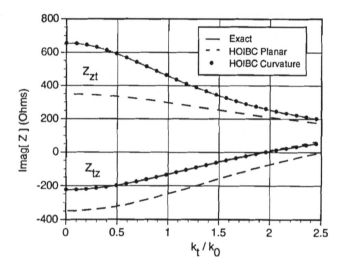

Figure 4.14: Impedance terms for a curved dielectric coating with $a = 0.1\lambda_0$, $d = 0.075\lambda_0$, $\epsilon_r = 6.0$, and $\mu_r = 1.0$.

correct impedance behavior for such a small radius of curvature, and including curvature effects is essential in predicting the correct RCS in this case.

In Chapter 6 both the planar HOIBC and the HOIBC that includes curvature effects are applied to two-dimensional dielectric-coated conducting bodies of arbitrary shape. Therein the improved accuracy of the HOIBC that include curvature will be illustrated further. Next a few examples are presented to illustrate the application of the HOIBC that include curvature effects to chiral coated circular cylinders.

As a first chiral example we consider the case of a coated cylinder with conductor radius $a = 0.9\lambda_0$, outer coating radius $b = 1.0\lambda_0$, and coating parameters $\epsilon_r = 4.0$, $\mu_r = 1.0$, and $\gamma_c = 0.003\Omega^{-1}$. This is the same coating considered previously, in the first planar chiral example of Chapter 3. Figure 4.15 shows the impedance terms versus the normalized wavenumber k_t/k_0. The exact impedances are shown along with those obtained using the HOIBC that includes

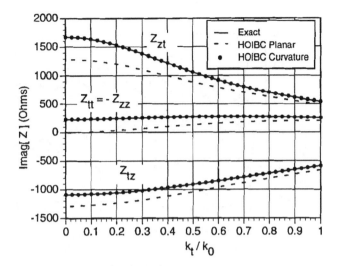

Figure 4.15: Impedance terms for a curved chiral coating with $a = 0.9\lambda_0$, $d = 0.1\lambda_0$, $\epsilon_r = 4.0$, $\mu_r = 1.0$, and $\gamma_c = 0.003\Omega^{-1}$.

curvature effects. Also included are the impedances obtained using the planar HOIBC. The HOIBC that includes curvature does an excellent job of modeling the behavior of the coating. All of the impedance terms are strongly affected by the curvature and are quite dissimilar to the impedances obtained using a locally planar approximation. The bistatic radar cross section for an H polarized wave incident in the $\phi = 0$ degree direction is plotted in Figure 4.16. An exact solution and HOIBC solutions which include and neglect curvature are shown. Excellent agreement between the HOIBC solution that includes the effects of curvature and the exact solution is seen, while the planar HOIBC results are unacceptable. The planar approximation is once again incapable of correctly computing the magnitudes of some of the higher order terms, those significantly excited. This example indicates that the effects of curvature must be included in the HOIBC in order to obtain a good approximation to the RCS for this particular choice of coating parameters.

As expected, the effects of curvature are less significant for larger cylinders. Figure 4.17 plots the H polarized RCS for a larger cylinder, $a = 4.9\lambda_0$ and $d = 0.1\lambda_0$, with the same coating as the previous example. In this case the planar approximation is valid, and the HOIBC including curvature gives only a slight improvement over that based on the planar approximation.

As an example of a small cylinder the case of a coated cylinder with conductor radius $a = 0.15\lambda_0$, outer coating radius $b = 0.25\lambda_0$, and coating parameters $\epsilon_r = 4.0$, $\mu_r = 1.0$, and $\gamma_c = 0.003\Omega^{-1}$ is considered in Figure 4.18. The RCS for H polarization incident is plotted. This example illustrates that the HOIBC that includes curvature predicts the co-polarized and cross-polarized

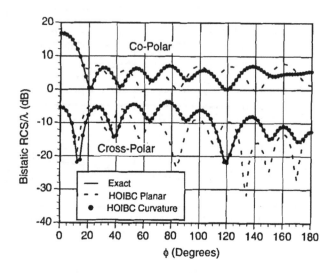

Figure 4.16: Bistatic radar cross section for a chiral coated cylinder with $a = 0.9\lambda_0$, $d = 0.1\lambda_0$, $\epsilon_r = 4.0$, $\mu_r = 1.0$, $\gamma_c = 0.003\Omega^{-1}$, and H polarization. Backscatter direction is 180 degrees.

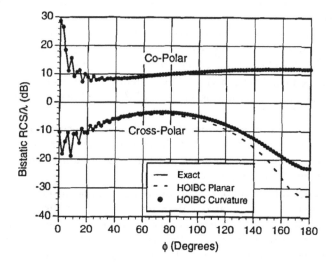

Figure 4.17: Bistatic radar cross section for a chiral coated cylinder with $a = 4.9\lambda_0$, outer coating radius $d = 0.1\lambda_0$, $\epsilon_r = 4.0$, $\mu_r = 1.0$, $\gamma_c = 0.003\Omega^{-1}$, and H polarization. Backscatter direction is 180 degrees.

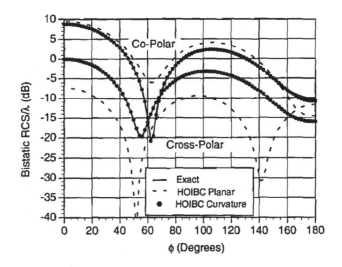

Figure 4.18: Bistatic radar cross section for a chiral coated cylinder with $a = 0.15\lambda_0$, $d = 0.1\lambda_0$, $\epsilon_r = 4.0$, $\mu_r = 1.0$, $\gamma_c = 0.003\Omega^{-1}$, and H polarization. Backscatter direction is 180 degrees.

components quite well even for small radii of curvature, whereas the planar approximation is inaccurate, particularly for the cross-polarized component.

Figure 4.19 shows results for the E polarized case with conductor radius $a = 1.98\lambda_0$, outer coating radius $b = 2.0\lambda_0$, and coating parameters $\epsilon_r = 1.0$, $\mu_r = 2.0$, and $\gamma_c = 0.005\Omega^{-1}$. For this large cylinder and thin coating both the planar HOIBC and HOIBC that includes curvature are accurate in the prediction of both components of the scattered field. The results for the H polarized case show similar agreement.

4.3   Conclusions

Application of the planar HOIBC for coated circular cylinders has demonstrated their ability to accurately predict the scattering properties of these objects, particularly when their radius of curvature is greater than one free space wavelength. For smaller cylinders curvature effects were found to be significant, and the planar HOIBC must be modified to include curvature effects. The cylinder examples have shown that a second order boundary condition that includes the effects of curvature correctly predicts the scattering cross section in a wide variety of circumstances. Examples were presented for simple dielectric coatings, as well as chiral coatings. It should also be reiterated that solutions based on a normally incident field approximation, such as the SIBC and TIBC, would fail to predict any cross polarized component of the reflected field for the case of

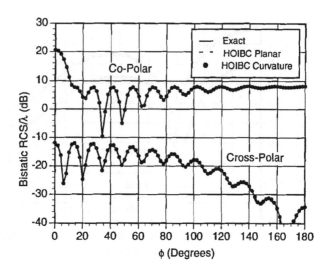

Figure 4.19: Bistatic radar cross section for a chiral coated cylinder with $a = 1.98\lambda_0$, $d = 0.02\lambda_0$, $\epsilon_r = 1.0$, $\mu_r = 2.0$, $\gamma_c = 0.005\Omega^{-1}$, and E polarization. Backscatter direction is 180 degrees.

a chiral-coated scatterer, whereas the HOIBC solutions are capable of modeling the complete behavior of the chiral coating. The next several chapters will employ the HOIBC in the solution of several practical scattering problems.

Chapter 5

Scattering by a Dielectric-Filled Groove in a Ground Plane

In this chapter the first of several examples of the application of higher order impedance boundary conditions to practical problems is considered. The scattering object to be considered in this chapter, a two-dimensional dielectric-filled groove in an infinite ground plane, is depicted in Figure 5.1. This object can be used to model a perturbation in an otherwise homogeneous large conducting body, such as an aircraft. Examples of these perturbations consist of hatches, ports, seams between panels, and cavity-backed antennas. The structure is assumed to be uniform in the z direction, and a plane wave incident at an angle ϕ_0, with $\mathbf{H}$ polarized in the z direction, is assumed. The conducting boundary of the groove is of arbitrary shape, and the dielectric filling the groove is assumed to be comprised of discrete layers oriented parallel to the ground plane, as depicted in Figure 5.1.

Scattering by single layer grooves with a lossy dielectric have been considered recently in Ref. [39], using the Standard Impedance Boundary Condition. Third order generalized impedance boundary conditions have been applied to the problem of single layer rectangular and tapered grooves in Ref. [26]. In this chapter higher order impedance boundary conditions that are determined

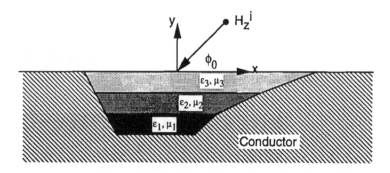

Figure 5.1: A dielectric-filled groove in an infinite ground plane.

using the methods of Chapter 2 are applied to the problem of scattering by multilayer grooves with arbitrary profiles, and the effects of discontinuities on the boundary conditions, i.e., sharp corners, is addressed. The monostatic and bistatic radar cross section of the groove will be computed using higher order impedance boundary conditions as well as the Standard Impedance Boundary Condition. These results will be compared to an exact solution that is based upon modeling the groove as a connected set of parallel plate waveguides terminated in a short. The approximate solution method as well as the exact method will be described in the next section. The final section will present results for various groove configurations.

5.1   Higher Order Impedance Boundary Condition Solution

5.1.1   Exterior Region

The higher order impedance boundary condition solution of the scattering problem begins by defining two equivalent problems, one for the exterior region, $y > 0$, and another for the interior region, $y < 0$. For the exterior region the materials below $y = 0^+$ are replaced by equivalent magnetic and electric currents, $\mathbf{M}$ and $\mathbf{J}$, which produce the true scattered fields in the region $y > 0$, as depicted in Figure 5.2a and b. These equivalent currents are given by

$$\mathbf{J} = \hat{\mathbf{y}} \times \mathbf{H} \tag{5.1}$$

and

$$\mathbf{M} = -\hat{\mathbf{y}} \times \mathbf{E}. \tag{5.2}$$

Due to the presence of the perfect conductor the equivalent magnetic current exists only in the groove aperture, whereas the equivalent electric current is present for all x. Since the equivalent sources and incident fields produce zero

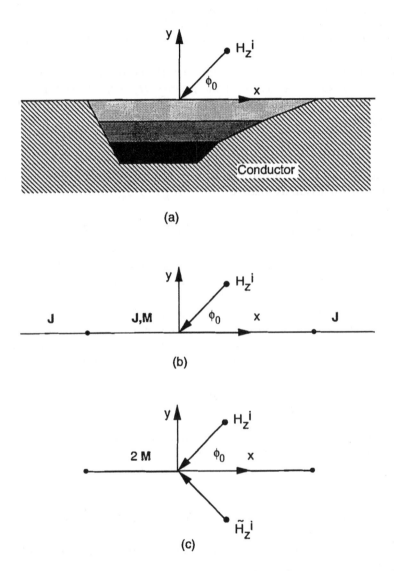

Figure 5.2: Derivation of the exact equivalent problem for the exterior region.

total field in the region $y < 0$, a conducting ground plane may be placed immediately below them, shorting out the electric current, **J**. The effect of the ground plane on the magnetic current is accounted for by using image theory, resulting in a doubled magnetic current in the aperture, as depicted in Figure 5.2c. The effect of the ground plane on the incident field, H_z^i, is accounted for by including an image of the incident field, $\tilde{H}_z^i$, as depicted in the figure.

The quantity of interest in the solution of the problem is the total tangential magnetic field in the aperture of the groove, $H_z^{Tot}(x)$. It is a result of the incident field, H_z^i, the image incident field, $\tilde{H}_z^i$, and the scattered field due to the magnetic current, $H_z^S(x)$,

$$H_z^{Tot}(x) = H_z^i(x) + \tilde{H}_z^i(x) + H_z^S(x). \tag{5.3}$$

The scattered field, H_z^S, is determined from the magnetic sources using the free space Green's function as follows,

$$H_z^S(x) = -\frac{\omega\epsilon_0}{2} \int_{x_l}^{x_r} M_z(x') H_0^{(2)}(k|x - x'|) dx', \tag{5.4}$$

where $H_0^{(2)}$ is the Hankel function of the second kind, zeroth order, and x_l and x_r are the left and right edges of the groove in the $y = 0$ plane.

Equation (5.4) is the first equation relating the magnetic field and electric field, M_x, in the aperture of the groove. It is applicable to the exact solution of the problem as well as to the higher order impedance boundary condition solution. A second equation relating the two quantities that takes into account the effects of the groove, i.e., the region $y < 0$, is required to complete the solution.

5.1.2 Interior Region

The higher order impedance boundary condition solution proceeds by assuming that the total electric and magnetic fields in the aperture are related to each other through a second order differential equation,

$$E_x^{Tot}(x) - c_2(x)\frac{\partial^2 E_x^{Tot}(x)}{\partial x^2} = c_0(x)H_z^{Tot}(x) - c_1(x)\frac{\partial^2 H_z^{Tot}(x)}{\partial x^2}. \tag{5.5}$$

Using Eq. (5.2) the boundary condition equation, Eq. (5.5), may be written in terms of the magnetic current $M_z(x)$,

$$M_z^{Tot}(x) - c_2(x)\frac{\partial^2 M_z^{Tot}(x)}{\partial x^2} = c_0(x)H_z^{Tot}(x) - c_1(x)\frac{\partial^2 H_z^{Tot}(x)}{\partial x^2}. \tag{5.6}$$

Here the constants $c_0(x)$, $c_1(x)$, and $c_2(x)$ are known and depend on the characteristics of the groove immediately below the point x. The method for the

determination of these coefficients for the multilayer dielectric follows directly
from that described in Chapter 2. For each position x an equivalent canonical
problem is constructed as depicted in Figure 5.3. Assuming a propagating

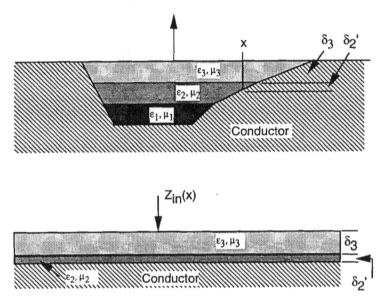

Figure 5.3: Determination of the local input impedance for each position in the
aperture of the groove.

wave of the form $e^{-jk_x x}$ the input impedance at $y = 0$ may be obtained by
transferring the short circuit through the multiple layers of material using the
standard transmission line equation,

$$Z_{in}^i = Z_0^i \frac{Z_{in}^{i-1} + jZ_0^i \tan k_y^i \delta_i}{Z_0^i + jZ_{in}^{i-1} \tan k_y^i \delta_i}, \tag{5.7}$$

with

$$k_y^i = \sqrt{\omega^2 \mu_i \epsilon_i - k_x^2} \tag{5.8}$$

and

$$Z_0^i = \sqrt{\frac{\mu_i}{\epsilon_i}} \frac{k_y^i}{k_i}, \tag{5.9}$$

with $k_i = \omega\sqrt{\mu_i \epsilon_i}$. For the particular position x depicted in Figure 5.3 the initial
short circuit is located in the (ϵ_2, μ_2) layer. The short is cascaded a distance
δ_2' through this layer using Eq. (5.7), and finally through the (ϵ_3, μ_3) layer to
obtain the final input impedance $Z_{in}|_x(k_x)$.

The appropriate coefficients for each position x are determined by matching
the input impedance with a ratio of polynomial approximation in a manner

identical to that presented in Chapter 2,

$$\frac{c_0(x) + c_1(x)k_x^2}{1 + c_2(x)k_x^2} \approx Z_{in}|_z(k_x) \tag{5.10}$$

All three coefficients may be determined by matching the input impedance at each position x for $k_x = 0$, $k_x = k_0/2$, and $k_x = k_0$. Various other options are available. For example, if c_1 and c_2 are set equal to zero and c_0 is chosen to be equal to the input impedance for $k_x = 0$, the Standard Impedance Boundary Condition is obtained. Boundary conditions of intermediate complexity are obtained by choosing either c_1 or c_2 equal to zero and using one additional match point in addition to $k_x = 0$.

Having determined the coefficients the overall equation to be solved by substituting (5.4) into (5.6) and rearranging,

$$\left[1 + c_2(x)\frac{\partial^2}{\partial x^2}\right] M_z(x)$$

$$+ \left[c_0(x) + c_1(x)\frac{\partial^2}{\partial x^2}\right] \frac{\omega\epsilon_0}{2} \int_{x_l}^{x_r} M_z(x')H_0^{(2)}(k_0|x - x'|)dx'$$

$$= 2\left[c_0(x) + c_1(x)\frac{\partial^2}{\partial x^2}\right] e^{jk_0\cos\phi_0 x}, \tag{5.11}$$

where the expressions for the incident field and its image have been used,

$$H_z^i(x) = \tilde{H}_z^i(x) = e^{jk_0\cos\phi_0 x}. \tag{5.12}$$

Equation (5.11) is solved using the method of moments, expanding the unknown function $M_z(x)$ in terms of pulse functions,

$$M_z(x) = \sum_{i=1}^{N_p} M_i P_i(x), \tag{5.13}$$

and using Dirac delta functions located at the center of each pulse, $\delta(x - x_j)$, as the testing functions, forcing Eq. (5.11) to be satisfied exactly at these points. If N_p such functions are used to expand $M_z(x)$, then the width of each pulse is $\Delta = (x_r - x_l)/N_p$ and the center of the ith pulse is located at $x_i = x_l + (i - 1/2)\Delta$. The second derivatives are approximated using a finite difference formula,

$$\frac{\partial^2 F(x_i)}{\partial x^2} \approx \frac{F(x_{i+1}) - 2F(x_i) + F(x_{i-1})}{\Delta^2}. \tag{5.14}$$

The final matrix equation to be solved for the unknown coefficients M_i is then given by

$$[Y]\{M\} = \{I\}. \tag{5.15}$$

The elements of the matrix $[Y]$ are given by,

$$Y(i,j) = \frac{\omega\epsilon_0}{2}\left[\left(c_0(x_i) - \frac{2c_1(x_i)}{\Delta^2}\right)\right.$$
$$\left.\cdot\left(\Delta\left(1 - \frac{2j}{\pi}\log\left(\frac{\gamma k_0\Delta}{4e}\right)\right)\right) + \frac{2c_1(x_i)H_0^{(2)}(k_0\Delta)}{\Delta}\right]$$
$$+ \left(1 - \frac{2c_2(x_i)}{\Delta^2}\right); \quad i = j, \tag{5.16}$$

$$Y(i,j) = \frac{\omega\epsilon_0}{2}\left[c_0(x_i)\Delta H_0^{(2)}(k_0\Delta) + \frac{2c_1(x_i)}{\Delta^2}\left(\Delta\left(1 - \frac{2j}{\pi}\log\left(\frac{\gamma k_0\Delta}{4e}\right)\right)\right.\right.$$
$$\left.\left.+ \Delta H_0^{(2)}(2k_0\Delta) - 2\Delta H_0^{(2)}(k_0\Delta)\right)\right] + \frac{c_2(x_i)}{\Delta^2}; \quad i = j \pm 1, \tag{5.17}$$

and

$$Y(i,j) = \frac{\omega\epsilon_0}{2}\left[\left(c_0(x_i) - \frac{2c_1(x_i)}{\Delta^2}\right)H_0^{(2)}(k_0|x_i - x_j|)\right.$$
$$\left.+ \frac{c_1(x_i)}{\Delta^2}\left(H_0^{(2)}(k_0|x_i - x_j + \Delta|) + H_0^{(2)}(k_0|x_i - x_j - \Delta|)\right)\right];$$
$$|i - j| > 1, \tag{5.18}$$

where $\gamma = 1.781$, Euler's constant arising in the small argument approximation for the Hankel function. The elements of the source vector are given by

$$I(i) = 2\left(c_0(x_i) - c_1(x_i)k_0^2\cos^2\phi_0\right)e^{jk_0\cos\phi_0 x_i}. \tag{5.19}$$

5.1.3 Edge Conditions

Although it would appear that the HOIBC solution of the problem is now complete, this is not the case. Some supplemental conditions need to be included at the ends of the groove. This is apparent from examination of Eq. (5.14). In order to properly evaluate the second derivatives M_z at the two ends of the groove, $i = 1$ and $i = N_p$, the values of $M_z(x_l - \Delta/2)$ and $M_z(x_r + \Delta/2)$ are required. Since these positions lie outside the aperture of the groove, M_z is not expanded there, and hence is undefined. In the implementation presented in Ref. [26], where a conjugate gradient-Fast Fourier Transform (FFT) method is used to solve the HOIBC equations, the unknown vector, M_z, is padded with zeros outside the actual space occupied by the groove aperture. Therefore the values of $M_z(x_l - \Delta/2)$ and $M_z(x_r + \Delta/2)$ are implicitly set to zero. As will be seen in the examples section to follow, this has the detrimental effect of driving M_z to zero at the edges of the groove. In general M_z exhibits edge behavior except when the groove depth tapers to zero gradually at both ends.

These effects are demonstrated quite clearly in Figure 4a, of Ref. [26], where M_z for a rectangular groove is shown, The exact solution exhibits edge behavior, whereas the GIBC solution incorrectly drives M_z to zero. In the case of a tapered groove, Figure 7 of Ref. [26], M_z is indeed tending toward zero at the ends of the groove due to the decreasing depth, and padding M_z with zeros outside the groove does not degrade the solution significantly.

The fact that the HOIBC must be supplemented with other conditions at discontinuities has been pointed out in Ref. [31]. For the groove problem the appropriate supplemental conditions are edge conditions. In Figure 5.4 the aperture of the groove is divided into three regions, Δ_r, Δ_m, and Δ_l. The regions Δ_r and Δ_l at the right and left ends of the groove represent regions where edge conditions are applied in lieu of the HOIBC. The middle region Δ_m is the region where the HOIBC is applied. As the regions Δ_r and Δ_l tend toward zero the edge condition applies more rigorously, whereas the region Δ_m expands toward the edges of the groove. As Δ_m expands the HOIBC becomes less and less accurate at its edges. Thus, a balance must be struck between the size of the edge condition regions and the HOIBC regions. This will be discussed in greater detail in the examples section.

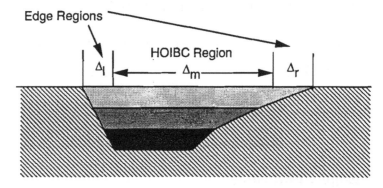

Figure 5.4: Division of the groove aperture into regions where HOIBC and edge conditions are enforced.

The appropriate edge conditions to be applied at the edges of the groove are derived from the canonical problem of scattering at a perfectly conducting wedge embedded in a number of wedges of dielectric material [40, 41], as depicted in Figure 5.5. For the groove problem we are concerned with the case of a conducting wedge with angle Φ_0, and two dielectric regions, the first of which represents the outermost dielectric layer in the groove, Φ_1, and the second of which represents the free space region, Φ_2. Following the development in Ref. [40], the electric field normal to the edge has the behavior

$$E_\rho \sim \rho^{t-1}, \tag{5.20}$$

where ρ is the distance from the edge and t is the eigenvalue that determines

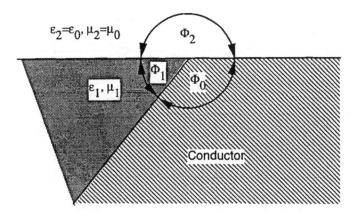

Figure 5.5: Canonical problem for determination of edge conditions in the groove aperture.

the leading behavior of the field. The value of t is determined by solving the following equation [40]

$$\frac{\tan t\Phi_1}{\epsilon_1} + \frac{\tan t\Phi_2}{\epsilon_2} = 0. \tag{5.21}$$

As the value of t approaches 1 the field becomes less singular, and when t approaches 0 the field varies as $1/\rho$. For the case of a groove with square corners $\Phi_0 = 90$ degrees, $\Phi_1 = 90$ degrees, and $\Phi_2 = 180$ degrees. In all cases $\epsilon_2 = \epsilon_0$. Figure 5.6 plots the value of t as a function of ϵ_r, when $\epsilon_1 = \epsilon_r\epsilon_0$, for several values of Φ_1. As can be seen in the figure, as ϵ_r increases t tends to increase and the field becomes less singular. Smaller corner angles also result in less singular fields, i.e., higher values of t. Figure 5.7 plots the value of t versus the angle of the dielectric wedge, Φ_1 for $\epsilon_r = 2$, $\epsilon_r = 4$, and $\epsilon_r = 8$. The case $\Phi_1 = 0$ degrees corresponds to the limit of an infinitely long taper section, whereas as Φ_1 approaches 90 degrees the groove terminates in a sharp rectangular corner. The figure indicates that as Φ_1 decreases the field becomes less singular, further confirming the fact that reasonable results may be expected for highly tapered grooves, even when the edge condition is ignored.

The edge condition may be implemented in a number of ways. For this particular problem a very simple implementation has been selected. A number of unknowns on each end of the groove aperture are designated as edge pulses. Their magnitudes are derived through the edge condition and the nearest non-edge pulse. For example, if a single edge pulse is used at each end of the groove aperture, instead of satisfying the HOIBC at $x = x_l + \Delta/2$, the magnitude of the first current pulse is derived from the second according to

$$\frac{M_1}{\left(\frac{\Delta}{2}\right)^{t_l-1}} = \frac{M_2}{\left(\frac{3\Delta}{2}\right)^{t_l-1}}, \tag{5.22}$$

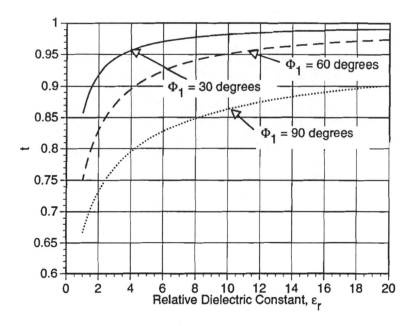

Figure 5.6: The effect of relative dielectric constant, ϵ_r, on the value of t for 90, 60, and 30 degree corners.

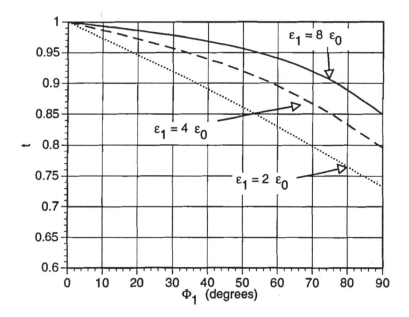

Figure 5.7: The effect of wedge angle, Φ_1, on the value of t for $\epsilon_r = 2$, 4, and 8.

where t_l is the value of t associated with the left edge of the groove. Likewise for the right edge of the groove,

$$\frac{M_{N_p}}{\left(\frac{\Delta}{2}\right)^{t_r-1}} = \frac{M_{N_p-1}}{\left(\frac{3\Delta}{2}\right)^{t_r-1}}, \tag{5.23}$$

where t_r is associated with the groove's right edge. This equation replaces the equation enforcing the HOIBC at $x = x_r - \Delta/2$. It should be noted that t_r and t_l are not equal unless the groove is tapered symmetrically. These results are easily extended to the situation when more than one edge pulse is used at a groove edge. The effect of including edge pulses will be demonstrated through several examples, in a later section, but first an exact solution of the scattering problem will be discussed.

5.2 Exact Formulation: A Mode Matching Approach

In order to obtain an exact solution to the problem of scattering by the tapered dielectric-filled groove, one may employ a finite element solution as discussed in Ref. [26]. If the groove is rectangular and filled with a homogeneous dielectric a simple modal-based solution is possible, also discussed in Ref. [26]. In this section an exact solution to the problem at hand is obtained by extending the modal solution to the case of arbitrary shaped grooves with layered dielectrics, avoiding the finite element method.

The equivalent problem for the exterior region $y > 0$ is identical to that depicted in Figure 5.2. As before, the total magnetic field and electric fields in the aperture of the groove are related through Eq. (5.3).

As with the HOIBC solution a pulse expansion is used for the magnetic current, substituted into Eq. (5.3), and the magnetic field is evaluated at the center of each pulse, resulting in the following matrix equation,

$$[Y_1]\{M\} - \{I\} = \{H_z\}. \tag{5.24}$$

Here the elements of $\{I\}$ are given by Eq. (5.19), and the elements of $[Y_1]$ are given by

$$Y_1(i,j) = \frac{\omega\epsilon_0\Delta}{2}\left(1 - \frac{2j}{\pi}\log\left(\frac{\gamma k_0\Delta}{4e}\right)\right); \; i = j, \tag{5.25}$$

and

$$Y(i,j) = \frac{\omega\epsilon_0\Delta}{2}H_0^{(2)}(k_0|x_i - x_j|); \; i \neq j \tag{5.26}$$

and $H_z(i)$ represents the value of H_z at position x_i.

An additional equation relating the tangential electric field M and magnetic field H_z components in the aperture of the groove is obtained by considering the interior problem depicted in Figure 5.8. The boundary of the groove is

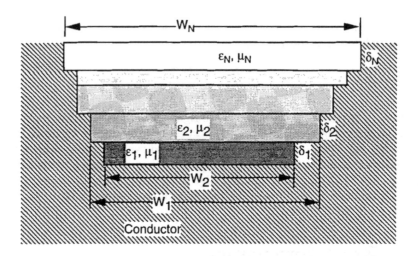

Figure 5.8: Equivalent exact problem for the interior region.

approximated by a series of small steps. The groove is thus approximated as a number of parallel plate waveguide regions, each of width w, connected in series. The width w, depth d, and material properties ϵ_r and μ_r of each region are in general different. As the number of segments used to approximate the groove increases the errors introduced by the stairstep approximation decrease, usually reaching acceptable levels for a approximately 10 steps per wavelength of depth. Given the geometry of Figure 5.8 an additional equation that relates the tangential field in the groove aperture to the electric field may be obtained through the use of mode matching methods [42]. The mode matching solution approaches the exact solution, subject to the geometrical approximations discussed earlier, as the number of modes used to expand the field in each region approaches infinity. In practice, convergence of the solution is obtained by including a modest number of modes. A detailed explanation of the mode matching solution follows in the next few pages.

We assume that the electric field, or equivalently the magnetic current in the aperture is again approximated by a set of pulse functions. The goal of the present development is then to determine a relationship between the magnetic field at any point in the aperture due to a pulse of electric field in the aperture. Subject to the step approximation of Figure 5.8 the fields at the aperture may be expressed in terms of a set of forward and reverse traveling waveguide modes appropriate for the uppermost waveguide segment, denoted as waveguide number N_s. If $a(i)$ denotes the unknown amplitude of the ith $-y$ traveling mode, $b(i)$ the amplitude of the corresponding y traveling mode, $e_i(x)$ is the x component of the electric field for the ith mode and $h_i(x)$ is the z component of the magnetic field then the total tangential fields in the aperture may be written as a sum as follows

$$E_z^{N_s}(x) = \sum_{i=1}^{I} (a^{N_s}(i) + b^{N_s}(i)) e_i^{N_s}(x) \tag{5.27}$$

and

$$H_z^{N_s}(x) = \sum_{i=1}^{I} (a^{N_s}(i) - b^{N_s}(i)) h_i^{N_s}(x), \tag{5.28}$$

where I modes have been used to represent the total field in the waveguide region. The appropriate modes for the parallel plate waveguide are the TE to z modes, and $e_i(x)$ and $h_i(x)$ are then given by,

$$e_i(x) = \sqrt{\frac{Z_i \delta_i}{w}} \cos{(k_i(x - x_l))}, \tag{5.29}$$

$$h_i(x) = \frac{e_i(x)}{Z_i}, \tag{5.30}$$

with

$$k_i = \frac{(i-1)\pi}{w}, \tag{5.31}$$

$$Z_i = \frac{\gamma_i}{j\omega\epsilon}, \tag{5.32}$$

$$\delta_i = \begin{cases} 2, & \text{if } i = 1; \\ 1, & \text{otherwise}, \end{cases} \tag{5.33}$$

and

$$\gamma_i = \sqrt{k_i^2 - k_2}. \tag{5.34}$$

Here x_l, k, ϵ, μ, and w are appropriate for the parallel plate region of interest. The normalization of the field components is such that

$$\int_{x_l}^{x_r} e_i(x) h_j(x) dx = \begin{cases} 1, & \text{if } i = j; \\ 0, & \text{otherwise}. \end{cases} \tag{5.35}$$

The first step in determining the magnetic field due to a pulse shaped distribution of electric field is to represent the pulse in terms of the modal expansion,

$$P_k(x) = \sum_{i=1}^{I} (a_k(i) + b_k(i)) e_i(x) \tag{5.36}$$

where $a_k(i)$ and $b_k(i)$ are the forward and reverse mode amplitudes due to the kth pulse. They may be determined by multiplying both sides of the equation by $e_i(x)$, integrating over the aperture of the groove, and making use of the orthogonality of the waveguide modes,

$$E(i,k) = (a_k(i) + b_k(i)) = \frac{\int_{x_l}^{x_r} P_k(x) e_i(x) dx}{\int_{x_l}^{x_r} e_i(x) e_i(x) dx}. \tag{5.37}$$

In order to determine the corresponding magnetic field, the composite scattering matrix for the series of parallel plate waveguides making up the groove is used. It relates the vectors $\{a\}$ and $\{b\}$,

$$\{b\} = [S]\{a\}. \tag{5.38}$$

The determination of the scattering matrix, $[S]$, will be outlined in subsequent pages. Using Eqs. (5.27) and (5.28), the vector $\{a_k\} - \{b_k\}$ is determined in terms of $\{a_k\} + \{b_k\}$,

$$\{a_k\} - \{b_k\} = \left[\tilde{Y}\right](\{a_k\} + \{b_k\}), \tag{5.39}$$

where

$$\left[\tilde{Y}\right] = ([I] - [S])([I] + [S])^{-1}. \tag{5.40}$$

Given these relationships it is a simple matter to find the magnetic field at the center of the mth pulse due to the kth pulse of electric field,

$$H_z(x_m) = \sum_{i=1}^{I}(a_k(i) - b_k(i))h_i(x_m) = \sum_{j=1}^{I}\sum_{i=1}^{I}h_i(x_m)\tilde{Y}(i,j)E(j,k). \tag{5.41}$$

If we denote $Y_2(m,k)$ as the magnetic field at the center of mth pulse due to the kth pulse of electric field, then

$$[Y_2] = [H]\left[\tilde{Y}\right][E], \tag{5.42}$$

with

$$H(m,j) = h_j(x_m), \tag{5.43}$$

and the elements of $\left[\tilde{Y}\right]$ are given by Eq. (5.40) in terms of the scattering matrix $[S]$.

The final matrix equation to be solved is then obtained by combining Eqs. (5.24) and (5.42),

$$[Y^{Tot}]\{M\} = ([Y_1] - [Y_2])\{M\} = \{I\}. \tag{5.44}$$

In order to complete the solution, the composite scattering matrix for the parallel plate waveguide approximation to the groove is required. The composite scattering matrix is obtained by cascading the matrices for each of the junctions making up the groove. An example of such a junction between two waveguides with different material properties and dimensions is shown in Figure 5.9. The first step in obtaining the scattering matrix for such a junction is to expand the tangential fields in each of the waveguides in terms of the appropriate modes,

$$E_x^{(1)}(x) = \sum_{i=1}^{N_1}(a^{(1)}(i) + b^{(1)}(i))e_i^{(1)}(x), \tag{5.45}$$

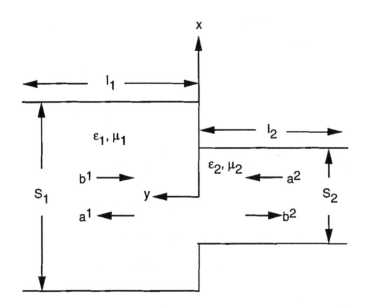

Figure 5.9: A junction between two parallel plate waveguides.

$$H_z^{(1)}(x) = \sum_{i=1}^{N_1}(a^{(1)}(i) - b^{(1)}(i))h_i^{(1)}(x), \tag{5.46}$$

$$E_x^{(2)}(x) = \sum_{i=1}^{N_2}(a^{(2)}(i) + b^{(2)}(i))e_i^{(2)}(x), \tag{5.47}$$

and

$$H_z^{(2)}(x) = \sum_{i=1}^{N_2}(a^{(2)}(i) - b^{(2)}(i))h_i^{(2)}(x). \tag{5.48}$$

The mode fields e and h are given by Eqs. (5.29) and (5.30) with the appropriate values for w, ϵ, and μ substituted for each region.

In order to determine the mode vectors reflected from the junction $\{a^{(1)}\}$ and $\{b^{(2)}\}$, in terms of the incident modes $\{a^{(2)}\}$ and $\{b^{(1)}\}$, the tangential fields are matched across the common interface at $y = 0$. Taking the product of the total electric fields with the magnetic field of each mode of waveguide 1 and integrating over the common region, S_2, gives

$$\int_{S_2} E_x^{(1)}(x)h_i^{(1)}(x)dS = \int_{S_2} E_x^{(2)}(x)h_i^{(1)}(x)dS. \tag{5.49}$$

Since $E^{(1)} = 0$ over the conductor surface $S_1 - S_2$, the left integral may be extended over S_1,

$$\int_{S_1} E_x^{(1)}(x)h_i^{(1)}(x)dS = \int_{S_2} E_x^{(2)}(x)h_i^{(1)}(x)dS. \tag{5.50}$$

Using the orthogonality of the mode vectors, Eq. (5.35), the preceding equation may be expressed in matrix form as

$$(\{a^{(1)}\} + \{b^{(1)}\}) = [P](\{a^{(2)}\} + \{b^{(2)}\}), \tag{5.51}$$

with

$$P(i,j) = \frac{1}{Z_i} \int_{S_2} e_i^{(1)}(x) e_j^{(2)}(x) dS. \tag{5.52}$$

Similarly, taking the product of the total magnetic fields with the electric field of each mode of waveguide 2 and integrating over the common region, S_2, gives

$$\int_{S_2} H_z^{(1)}(x) e_j^{(2)}(x) dS = \int_{S_2} H_z^{(2)}(x) e_j^{(2)}(x) dS. \tag{5.53}$$

Substituting in the expressions for the field vectors gives

$$(\{a^{(2)}\} + \{b^{(2)}\}) = [Q](\{a^{(1)}\} - \{b^{(1)}\}), \tag{5.54}$$

with

$$P(i,j) = Q(j,i), \tag{5.55}$$

or

$$[Q] = [P^T], \tag{5.56}$$

where $[P^T]$ indicates the transpose of $[P]$. The scattering matrix for the junction, referenced to $y = 0^+$ and $y = 0^-$, is defined by

$$\{a^{(1)}\} = [S'_{11}]\{b^{(1)}\} + [S'_{12}]\{a^{(2)}\} \tag{5.57}$$

and

$$\{b^{(2)}\} = [S'_{21}]\{b^{(1)}\} + [S'_{22}]\{a^{(2)}\}. \tag{5.58}$$

The scattering matrices may be found in terms of the matrix $[P]$ through simple algebraic manipulation,

$$[S'_{11}] = ([I] + [P][P^T])^{-1}([P][P^T] - [I]), \tag{5.59}$$

$$[S'_{12}] = 2([I] + [P][P^T])^{-1}[P], \tag{5.60}$$

$$[S'_{21}] = 2([I] + [P^T][P])^{-1}[P^T], \tag{5.61}$$

and

$$[S'_{22}] = ([I] + [P^T][P])^{-1}([I] - [P^T][P]). \tag{5.62}$$

The final step is to adjust the reference planes from $y = 0^{\pm}$ to the correct locations,

$$S_{11}(i,j) = S'_{11}(i,j) e^{(-j(\gamma_i^{(1)} + \gamma_j^{(1)})l_1)}, \tag{5.63}$$

$$S_{12}(i,j) = S'_{12}(i,j) e^{(-j(\gamma_i^{(1)} l_1 + \gamma_j^{(2)} l_2))}, \tag{5.64}$$

$$S_{21}(i,j) = S'_{21}(i,j)e^{(-j(\gamma_i^{(2)}l_2 + \gamma_j^{(1)}l_1))}, \tag{5.65}$$

and

$$S_{22}(i,j) = S'_{22}(i,j)e^{(-j(\gamma_i^{(2)} + \gamma_j^{(2)})l_2)}. \tag{5.66}$$

The process of cascading the individual scattering matrices making up the various sections of the groove in order to determine the overall input scattering matrix S_{11}^{N} is illustrated in Figure 5.10. The new S_{11} matrix, S_{11}^n, is required,

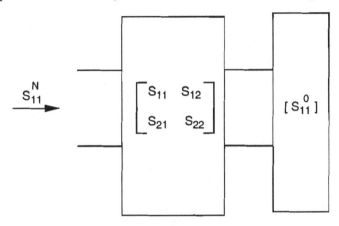

Figure 5.10: Cascading scattering matrices.

given the old scattering matrix S_{11}^{n-1} and the intervening device matrices. Again the process requires only algebraic manipulations,

$$[S_{11}^n] = [S_{11}] + [S_{12}]([I] - [S_{11}^{n-1}][S_{22}])^{-1}[S_{11}^{n-1}][S_{21}]. \tag{5.67}$$

The scattering matrix S_{11}^0 is initialized to $-[I]$, short circuit, at the bottom of the groove and then cascaded through each of the sections making up the groove according to the above equation, finally arriving at the overall scattering matrix S_{11}^{N}. This completes the details of the exact solution method.

5.3   Examples

In this section scattering by a number of example grooves will be examined using the exact formulation as well as the SIBC and HOIBC. Bistatic and monostatic RCS will be examined for single layer as well as multilayer grooves with straight and tapered edges. The effect of including edge behavior in the HOIBC formulation will be studied.

5.3.1   Rectangular Grooves

The first several examples will examine the ability of the HOIBC and SIBC formulations to predict the scattering properties of simple dielectric-filled rectangular grooves. We begin with an examination of the effects of groove depth on the accuracy of the approximate formulations. Figure 5.11 plots the monostatic RCS for a rectangular groove of length $3\lambda_0$ and depth $0.05\lambda_0$, filled with a dielectric, with $\epsilon_r = 5$. Results for the full range of angle of incidence are plotted, from grazing incidence, $\phi = 0$ degrees, to normal incidence, $\phi = 90$ degrees. In all of the examples of this section 15 pulses per free space wavelength were used to expand the magnetic field in the groove aperture. Results from the exact solution are compared to those based on the SIBC; a straightforward application of the HOIBC, as in Ref. [26], which ignores edge behavior; and the HOIBC in conjunction with a single edge pulse on each end of the groove, as discussed earlier in the chapter.

Figure 5.11 reveals that for near normal incidence, $\phi < 60$ degrees, all of the approximate formulations are in excellent agreement with the exact solution. As the incidence angle approaches grazing, $\phi = 0$ degrees, the SIBC and simple HOIBC results begin to deteriorate. When the edge behavior is added to the HOIBC solution excellent agreement with the exact solution is obtained for all angles of incidence. The trend of these results is consistent with expectations. The assumption of propagation normal to the boundary, which is central in the derivation of the SIBC, is quite valid for angles of incidence near normal, and hence the SIBC is quite accurate in this area. Since the HOIBC remedies the inaccuracy of the SIBC when there is a significant wavenumber along the boundary, both implementations of the HOIBC should have larger useful ranges than the SIBC. This will be demonstrated more dramatically in the following examples, as the depth of the groove increases. For angles of incidence near grazing, $\phi = 0$ degrees, the singular behavior of the fields at the edges of the groove is dominant in determining the scattered fields, and hence including the edge behavior to the HOIBC further extends the useful range of the HOIBC. The inclusion of edge pulses in the SIBC was found to be ineffective in improving the accuracy of the solution. The inclusion of edge pulses does nothing to correct the major fault of the SIBC, its inability to model effects due to propagation tangential to the boundary. In addition, increasing the number of edge pulses used in the HOIBC formulation beyond one produced decreased accuracy in all cases examined, and hence a single edge pulse is employed throughout the examples discussed in this section.

Figure 5.12 shows the bistatic RCS for the same groove, $d = 0.05\lambda_0$, $w = 3.0\lambda_0$, and $\epsilon_r = 5.0$, for an angle of incidence of $\phi = 30$ degrees. As for the monostatic case the HOIBC that includes the edge behavior is the most accurate, reproducing the exact solution over the entire angular range, followed by the simple HOIBC result, and finally by the SIBC result. The SIBC formulation shows significant deviation from the exact solution, even in the forward scatter direction, $\phi = 150$ degrees. Once again this is due to the inability of the SIBC to detect transverse propagation along the boundary and correct the impedance accordingly. The magnetic current corresponding to this case is plotted in Figure 5.13. Both HOIBC formulations produce excellent approximations to

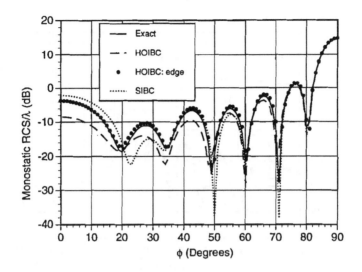

Figure 5.11: Monostatic RCS for a rectangular dielectric filled groove of width $3\lambda_0$ and depth $0.05\lambda_0$, with $\epsilon_r = 5.0$, $\mu_r = 1.0$, and H polarization.

the true magnetic current except near the edges of the groove. Near the edges of the groove the superiority of the HOIBC formulation that includes edge pulses over the simple formulation is clearly demonstrated. As seen in Figure 4 of Ref. [26], applying the HOIBC without including edge behavior incorrectly drives the magnetic current to zero at the ends of the groove. The SIBC current is seen to be quite inferior to the HOIBC currents throughout the groove aperture.

Figure 5.14 plots the monostatic RCS for the same groove as the previous examples when the depth of the groove is increased to $0.075\lambda_0$. The accuracy of the SIBC and the simple HOIBC solutions is reduced relative to those for a $0.05\lambda_0$ groove. For this groove the enhanced range of accuracy even for the simple HOIBC relative to the SIBC is apparent. As before the HOIBC solution that includes the edge behavior is quite accurate for all angles of incidence.

The next two figures present the final results for the rectangular groove with $\epsilon_r = 5.0$, this time for a depth of $0.1\lambda_0$. The monostatic RCS for the groove is plotted in Figure 5.15. In this case all the SIBC and simple HOIBC results are unacceptable, whereas the inclusion of a single edge pulse on each end of the groove produces excellent results. The bistatic RCS for $\phi = 30$ degrees is plotted in Figure 5.16. As can be seen in Figure 5.16 only the HOIBC formulations produce the correct sidelobe behavior, whereas the SIBC formulation gives only the average behavior.

A final lossless rectangular groove example is considered in Figure 5.17, where the monostatic RCS for a rectangular groove of width $3\lambda_0$ and depth $0.125\lambda_0$, filled with a dielectric with $\epsilon_r = 2.0$, is plotted. Only the HOIBC-edge

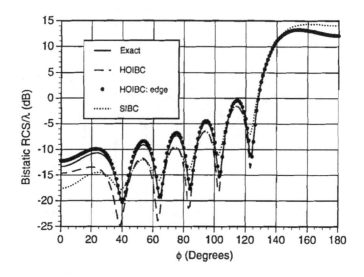

Figure 5.12: Bistatic RCS for a rectangular dielectric-filled groove of width $3\lambda_0$ and depth $0.05\lambda_0$, with $\epsilon_r = 5.0$, $\mu_r = 1.0$, $\phi_0 = 30$ degrees, and H polarization.

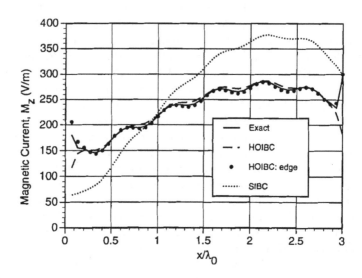

Figure 5.13: Magnetic current for a rectangular dielectric-filled groove of width $3\lambda_0$ and depth $0.05\lambda_0$, with $\epsilon_r = 5.0$, $\mu_r = 1.0$, $\phi_0 = 30$ degrees, and H polarization.

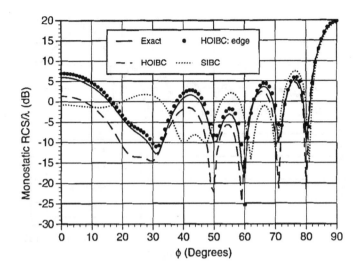

Figure 5.14: Monostatic RCS for a rectangular dielectric-filled groove of width $3\lambda_0$ and depth $0.075\lambda_0$, with $\epsilon_r = 5.0$, $\mu_r = 1.0$, and H polarization.

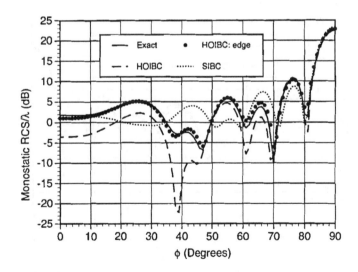

Figure 5.15: Monostatic RCS for a rectangular dielectric-filled groove of width $3\lambda_0$ and depth $0.1\lambda_0$, with $\epsilon_r = 5.0$, $\mu_r = 1.0$, and H polarization.

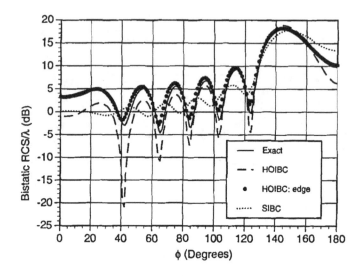

Figure 5.16: Bistatic RCS for a rectangular dielectric-filled groove of width $3\lambda_0$ and depth $0.1\lambda_0$, with $\epsilon_r = 5.0$, $\mu_r = 1.0$, $\phi_0 = 30$ degrees, and H polarization.

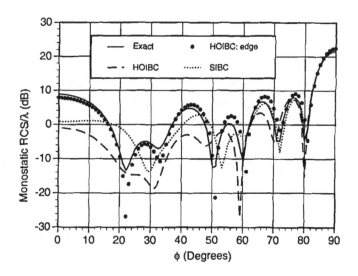

Figure 5.17: Bistatic RCS for a rectangular dielectric-filled groove of width $3\lambda_0$ and depth $0.125\lambda_0$, with $\epsilon_r = 2.0$, $\mu_r = 1.0$, and H polarization.

pulse formulation is capable of predicting the details of the scattering characteristics of this groove. The results presented in this subsection for lossless rectangular grooves are representative of a large number of cases examined. In all cases the HOIBC-edge pulse formulation produced acceptable results, with the best agreement with the exact solution obtained for grooves less than one quarter wavelength deep in the dielectric.

5.3.2 Rectangular Groove: Lossy Dielectric

It is well known that the SIBC formulation works well for grooves filled with lossy dielectric material, provided the groove depth is an appreciable fraction of a skin depth [39]. In this case the HOIBC formulation that includes the edge effects reverts to the SIBC solution. This is illustrated in Figure 5.18 where the monostatic RCS for a groove of width $3\lambda_0$, and depth $0.08\lambda_0$, filled with a lossy dielectric, $\epsilon_r = 5.0 - j5.0$, is plotted. In this case the depth of the groove corresponds to approximately one half skin depth. All three approximate formulations perform adequately, with the simple HOIBC result being least accurate. Examination of the current once again reveals that this is due to its inappropriately driving the current to zero at the ends of the groove. As the groove depth increases and/or the loss increases the HOIBC coefficients c_1 and c_2 tend to zero, indicating that the wave in the layer propagates essentially perpendicular to the boundary, as assumed in the SIBC. Thus the HOIBC, correctly applied, reverts to the SIBC in the lossy case. This was also demonstrated in Chapter 4, for scattering by conducting circular cylinders with lossy coatings. Finally, although the solution of Eq. (5.21) for complex ϵ_1 results in a complex value for t, the effect of the imaginary part of ϵ_r was found to produce only a small perturbation in the value of t, which was ignored. The results plotted in Figure 5.18 were obtained using the value of t corresponding to $\epsilon_r = 5.0$, i.e., $t = 0.81$.

5.3.3 Asymmetric Groove

Next scattering by a groove with a more complex geometry is considered. As an example the asymmetric groove depicted in Figure 5.19 is considered. The groove has a rectangular edge on the left end and a tapered edge on the right end. It is filled with a dielectric with $\epsilon_r = 3.0$ and reaches a depth of $0.1\lambda_0$. The overall length of the groove is $1.5\lambda_0$, with the taper section using a $0.5\lambda_0$ portion of the overall length. This represents a highly tapered end, with a wedge angle corresponding to $\Phi_1 = 11.3$ degrees. This example is chosen because of the dramatic difference in edge behavior at the two ends of the groove. The value of t at the rectangular edge, t_l, is 0.77, indicating singular behavior, whereas at the highly tapered right edge $t_r = 0.98$, indicating essentially no singularity in the current.

Due to the varying groove depth the HOIBC coefficients c_0 to c_2 now become functions of position in the groove aperture. This is illustrated in Figure 5.20,

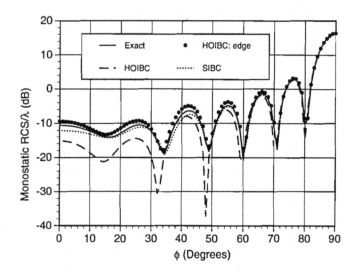

Figure 5.18: Monostatic RCS for a rectangular dielectric-filled groove of width $3\lambda_0$ and depth $0.08\lambda_0$, with $\epsilon_r = 5.0 - j5.0$, $\mu_r = 1.0$, and H polarization.

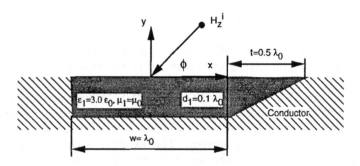

Figure 5.19: Geometry for a dielectric-filled groove with one tapered edge.

which plots the value of the coefficient c_0 as a function of position in the groove aperture. The value of the coefficient remains constant throughout the rectangular portion of the groove. The coefficient c_0, as well as the remaining coefficients, vary rapidly over the tapered portion of the groove, tending to zero at the end. The remaining coefficients c_1 and c_2 behave quite similarly and therefore are omitted from the plot.

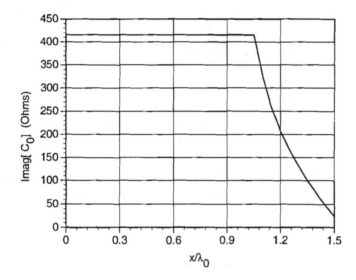

Figure 5.20: HOIBC coefficient, c_0 for a dielectric-filled groove with one tapered edge of overall width $1.5\lambda_0$, taper width $0.5\lambda_0$, and depth $0.1\lambda_0$, with $\epsilon_r = 3.0$, $\mu_r = 1.0$, and H polarization.

The monostatic RCS for this groove is depicted in Figure 5.21. Only the HOIBC solution that includes the edge pulses produces accurate results for all angles of incidence. The enhanced backscatter for grazing incidence on the sharp edge, $\phi = 0$ degrees, relative to that incident on the tapered edge, $\phi = 180$ degrees, is predicted well. As expected all of the approximate formulations perform well for angles of incidence near normal, $\phi = 90$ degrees.

In this nonrectangular case the exact solution consisted of modeling the tapered groove with 10 segments, expanding the field in each segment in terms of approximately 30 waveguide modes, and cascading scattering matrices for each of the discontinuities to obtain the final scattering matrix and overall solution. Since this process is quite time consuming, and the complexity of the HOIBC solution is only slightly increased relative to the rectangular case, the HOIBC represents an attractive alternative to the exact solution for this complex geometry. In this case the computation time for the HOIBC solution is less than one tenth that for the exact solution. The increased efficiency results in only a slight degradation of the accuracy of the solution, as seen in Figure 5.21.

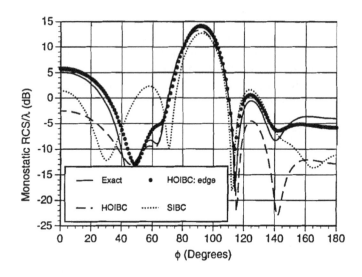

Figure 5.21: Monostatic RCS for a dielectric-filled groove with one tapered edge of overall width $1.5\lambda_0$, taper width $0.5\lambda_0$, and depth $0.1\lambda_0$, with $\epsilon_r = 3.0$, $\mu_r = 1.0$, and H polarization.

5.3.4   Tapered Groove: Two-Layer Dielectric

As a final example the tapered groove with a two-layer dielectric depicted in Figure 5.22 is chosen. In this case an 11.3 degree taper is included on both ends of the groove, and each dielectric region has a depth of $0.05\lambda_0$. This is a more complex example than that depicted in Figure 7 of Ref. [26], illustrating the versatility of the mode matching approach and the ability of the HOIBC to model complex structures relatively accurately.

The monostatic RCS of the groove is plotted in Figure 5.23. In this case both HOIBC formulations produce acceptable results whereas the SIBC is inadequate for predicting the groove's RCS. In this case the two HOIBC results are nearly identical since there is essentially no singularity at the groove edges, and assuming that the current tends to zero is valid. In this case the value of t at the edges of the groove is 0.99, indicating essentially no singularity. The same value for t would apply to the example of Figure 7 of Ref. [26], explaining the excellent results depicted in the figure.

The RCS for this groove when $\phi = 30$ degrees is shown in Figure 5.24, and the corresponding magnetic current, in Figure 5.25. Both HOIBC results are in good agreement with the exact solution, whereas the SIBC solution fails for essentially all ϕ, being inaccurate in both the backscatter as well as the

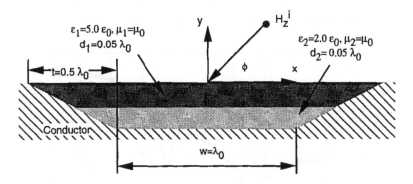

Figure 5.22: Geometry for a two-layer dielectric-filled groove with tapered edges.

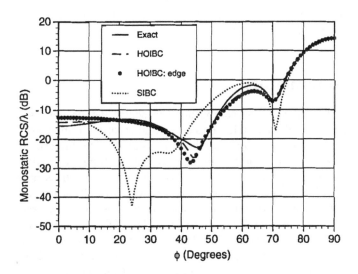

Figure 5.23: Monostatic RCS for the two-layer dielectric-filled groove depicted in Figure 5.22, and H polarization.

forward scatter directions. Figure 5.25 demonstrates the ability of the HOIBC to predict the correct aperture current, as well as the failure of the SIBC. The figure also illustrates the influence of the taper profile on the aperture current, forcing the current to zero at the ends of the groove.

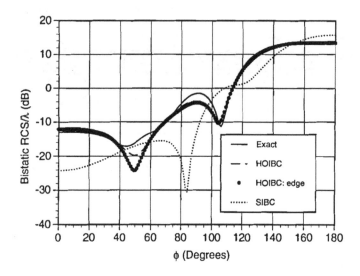

Figure 5.24: Bistatic RCS for the two-layer dielectric-filled groove depicted in Figure 5.22, and H polarization.

5.4   Conclusions

An efficient formulation for computing the scattering properties of tapered grooves with layered dielectrics based on higher order impedance boundary conditions and edge conditions has been presented. The bistatic and monostatic radar cross section for rectangular as well as tapered grooves has been computed. Single lossless and lossy dielectric layers as well as multilayer dielectric fillings have been considered. For grooves with either tapered or sharp corners results have compared quite favorably to those from an exact formulation based on a mode matching approach. The HOIBC solution requires less memory and less computation time than the exact solution, and is quite attractive for grooves with complex geometries. In general the HOIBC approach has been quite effective for groove depths of less than one quarter wavelength in the dielectric. As discussed in Ref. [26], for deeper grooves it is possible to use a hybrid HOIBC-exact approach, combining the advantages of both methods. The HOIBC that ignores edge effects was found to produce better results than the SIBC in essentially all cases. Including edge effects in the HOIBC formulation produces additional significant improvements as demonstrated by the

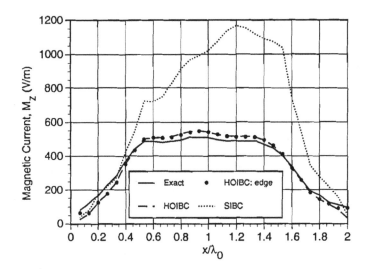

Figure 5.25: Magnetic current for the two-layer dielectric-filled groove depicted in Figure 5.22, and H polarization.

examples. The combination of HOIBC and edge conditions are thus capable of predicting the scattering properties of grooves for which the SIBC formulation is totally inadequate.

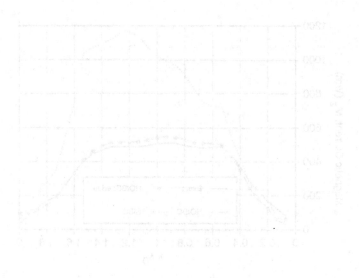

Chapter 6

Scattering by Two-Dimensional Dielectric-Coated Cylinders

In this chapter the methods discussed in Chapter 2 are applied to the problem of scattering by two-dimensional dielectric-coated cylinders [43]. Unlike the dielectric-filled groove problem discussed in the previous chapter, the present problem is inherently nonplanar. The accuracy of the boundary condition itself as well as the locally planar assumption are tested in this chapter. In addition, the boundary conditions that include the effects of curvature may be applied to this problem, demonstrating their increased accuracy relative to the planar boundary conditions. In order to evaluate these effects we study scattering of H polarized plane waves by two-dimensional coated conductors with external boundaries in the shape of a superquadric cylinder, Figure 6.1.

A number of bistatic and monostatic radar cross section examples will be presented to demonstrate that the higher order impedance boundary conditions can provide excellent results, far superior to those of the Standard Impedance Boundary Condition. When the effects of curvature are also taken into account, excellent results are obtained for bodies with relatively sharp corners.

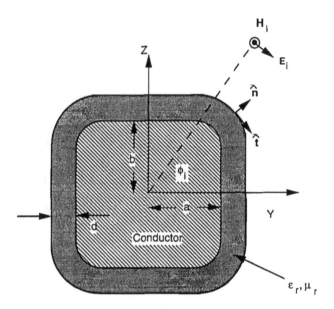

Figure 6.1: A general two-dimensional scattering problem (H polarization).

6.1 HOIBC Solution of the Scattering Problem

6.1.1 Exterior Problem

In this section the method of moments solution of the scattering problem is developed. Consider a two-dimensional scatterer where the relationship between the total electric and magnetic fields on the outer surface of the scatterer is described through an approximate impedance boundary condition that may involve differential operators. We begin by considering an equivalent exterior problem.

The total fields inside the body are defined as $\mathbf{E}_2\left(\mathbf{r}\right)$ and $\mathbf{H}_2\left(\mathbf{r}\right)$, while the total external fields are $\mathbf{E}_1\left(\mathbf{r}\right)$ and $\mathbf{H}_1\left(\mathbf{r}\right)$. The total external field is made up of both the incident and scattered fields,

$$\mathbf{E}_1\left(\mathbf{r}\right) = \mathbf{E}_1^s\left(\mathbf{r}\right) + \mathbf{E}_1^i\left(\mathbf{r}\right)$$
$$\mathbf{H}_1\left(\mathbf{r}\right) = \mathbf{H}_1^s\left(\mathbf{r}\right) + \mathbf{H}_1^i\left(\mathbf{r}\right). \tag{6.1}$$

To solve the scattering problem the body is replaced with equivalent electric and magnetic surface currents $\mathbf{M}_s$ and $\mathbf{J}_s$, which give the true field external to the body and zero field inside, i.e., $\mathbf{E}_2 = \mathbf{H}_2 = 0$. Using the definitions of the surface current $\mathbf{J}_s$ and $\mathbf{M}_s$ are found to be

$$
\begin{aligned}
\mathbf{J}_s &= \hat{n} \times \mathbf{H}_1, \\
\mathbf{M}_s &= -\hat{n} \times \mathbf{E}_1.
\end{aligned}
\tag{6.2}
$$

The sources for the scattered magnetic field $\mathbf{H}_1^s(\mathbf{r})$ are the electric and magnetic surface currents. The scattered field can be obtained from the magnetic and electric surface currents through

$$
\begin{aligned}
\mathbf{H}_1^s(\mathbf{r}) = \ & -j\omega\epsilon \oint_S \mathbf{M}_s(\mathbf{r}') G(\mathbf{r}, \mathbf{r}')\, ds \\
& -\frac{j\omega\epsilon}{k^2} \nabla \oint_S \nabla' \cdot \mathbf{M}_s(\mathbf{r}')\, G(\mathbf{r}, \mathbf{r}')\, ds \\
& +\nabla \times \oint_S \mathbf{J}_s(\mathbf{r}')\, G(\mathbf{r}, \mathbf{r}')\, ds,
\end{aligned}
\tag{6.3}
$$

where $\mathbf{r}$ is the observation point, $\mathbf{r}'$ is the source point, $\int_S$ indicates integration on the exterior surface of the body, and the two-dimensional free space Green's function is given by

$$
G(\mathbf{r}, \mathbf{r}') = \frac{1}{4j} H_0^2(k|\mathbf{r} - \mathbf{r}'|).
\tag{6.4}
$$

An analogous expression for the scattered electric field may be obtained from this expression using duality.

A Magnetic Field Integral Equation (MFIE) is obtained by satisfying

$$
\hat{n} \times \left(\mathbf{H}_1^s(\mathbf{r}) + \mathbf{H}_1^i(\mathbf{r}) \right) = 0,
\tag{6.5}
$$

where $\mathbf{r}$ is evaluated just inside the surface of the body.

Similarly an Electric Field Integral Equation (EFIE) is obtained by satisfying

$$
\hat{n} \times \left(\mathbf{E}_1^s(\mathbf{r}) + \mathbf{E}_1^i(\mathbf{r}) \right) = 0,
\tag{6.6}
$$

once again where $\mathbf{r}$ is evaluated just inside the surface of the body.

It is well known that both of the above formulations will fail at certain frequencies. These frequencies are the resonant frequencies of a conducting cavity having the same shape as the exterior surface of the coated body. This problem can be overcome by satisfying a linear combination of both equations, which is known as the Combined Field Integral Equation (CFIE).

6.1.2   Interior Problem

The next step in the formulation is to obtain a relationship between the unknown magnetic and electric surface currents on the body, thus reducing the number of unknowns by a factor of two. This is accomplished by solving the differential equation representing the boundary condition. For the two-dimensional scattering problem to be considered in this chapter, H polarization ($E_x = 0$) and $k_x = 0$ is assumed. In this case only the impedance $Z_{yx}(k_x, k_y)$, which was derived previously in Chapter 2, is of importance, implying the following exact boundary condition in the spectral domain for an infinite planar coating on a ground plane,

$$\tilde{E}_y(k_y) = j\sqrt{\frac{\mu_r\mu_0}{\epsilon_r\epsilon_0}}\frac{k_z}{k}\tan(k_z d)\,\tilde{H}_x(k_y). \tag{6.7}$$

Using this expression for the exact boundary condition in the spectral domain, the ratio of polynomial approximation for the impedance discussed in Chapter 2, and the properties of the Fourier transform, the following approximate boundary condition in the spatial domain is obtained

$$E_y(y) - c_2\frac{\partial^2 E_y(y)}{\partial y^2} = c_0 H_x(y) - c_1\frac{\partial^2 H_x(y)}{\partial y^2}. \tag{6.8}$$

In order to apply these boundary conditions on a general body the local coordinates shown in Figure 6.1 must be used. If "t" is defined as the distance along the contour describing the dielectric to free space interface, then the boundary condition becomes

$$E_t(t) - c_2(t)\frac{\partial^2 E_t(t)}{\partial t^2} = c_0(t)H_x(t) - c_1(t)\frac{\partial^2 H_x(t)}{\partial t^2}. \tag{6.9}$$

It should be noted that if the properties of the coating, such as thickness or dielectric constant, are a function of position on the body, then the constants c_0, c_1, and c_2 are also functions of "t" as is indicated in the equation.

When curvature is included in the boundary condition, Eq. (6.9) still applies, but the coefficients are now a function of the curvature of the body as well as the electrical parameters of the coating. Therefore, even for bodies with uniform coatings the coefficients will vary from point to point on the body due to changes in the radius of curvature along the boundary.

The application of the HOIBC that include curvature to the two-dimensional scattering problem is illustrated in Figure 6.2. The figure illustrates that when the curvature-based HOIBC are applied on the curved body the effects of the corners of the body are accounted for more accurately than for the planar HOIBC case. Since the curvature-based HOIBC reduce to the planar ones when the radius of curvature is large these improved boundary conditions should model the behavior of the coating more accurately in all situations. The examples that follow will illustrate that this is indeed the case for practical scattering by two-dimensional scattering objects.

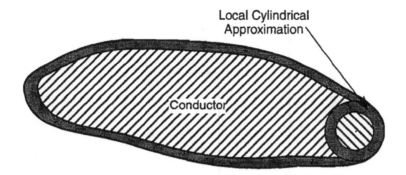

Figure 6.2: Transferring the Higher Order Impedance Boundary Condition that includes curvature onto the two-dimensional coated body.

6.1.3   Method of Moments Solution

First of all the method of moments is used to solve the boundary condition equation. Both the electric current and magnetic current are expanded in terms of spline functions, Figure 6.3. The spline function is chosen since both its first and second derivatives exist. The first derivative is a piecewise linear function, and the second derivative is represented by three pulses.

$$J_t(t) = \sum_{i=1}^{N} J_i S_i(t) \qquad (6.10)$$

$$M_x(t) = \sum_{i=1}^{N} M_i S_i(t). \qquad (6.11)$$

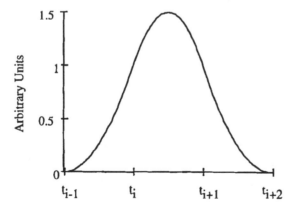

Figure 6.3: Spline expansion function.

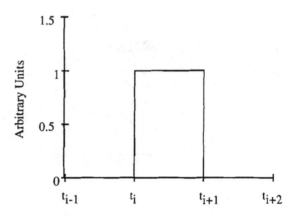

Figure 6.4: Pulse testing function.

Each spline function spans three subdivisions on the boundary of the scatterer and is represented as follows

$$S_i(t) = \frac{(t - t_{i-1})^2}{\Delta_1^2}, \, t_{i-1} < t < t_i, \tag{6.12}$$

$$S_i(t) = 1 + 2\frac{(t - t_i)}{\Delta_1} - \frac{(\Delta_1 + 2\Delta_2 + \Delta_3)(t - t_i)^2}{\Delta_1\Delta_2(\Delta_2 + \Delta_3)}, \, t_i < t < t_{i+1}, \tag{6.13}$$

$$S_i(t) = \frac{(\Delta_1 + \Delta_2)(t - t_{i+2})^2}{\Delta_1\Delta_3(\Delta_2 + \Delta_3)}, \, t_{i+1} < t < t_{i+2}. \tag{6.14}$$

with $\Delta_j = |t_{i+j-1} - t_{i+j-2}|$.

Employing the standard method of moments technique to solve the boundary condition equation and using pulse functions $P_j(t)$, Figure 6.4, for testing the equation, the following matrix equation is obtained

$$[B_M]\{M\} = [B_J]\{J\}. \tag{6.15}$$

The matrix elements are given by

$$B_M(i,j) = \oint_S P_i(t)\left[S_j(t) - c_2(t)\frac{\partial^2 S_j(t)}{\partial t^2}\right]dt \tag{6.16}$$

$$B_J(i,j) = \oint_S P_i(t)\left[c_0(t)S_j(t) - c_1(t)\frac{\partial^2 S_j(t)}{\partial t^2}\right]dt, \tag{6.17}$$

where $\oint$ indicates a line integral around the boundary of the scatterer.

One may solve for $\{M\}$ in terms of $\{J\}$ as follows

$$\{M\} = [B]\{J\}, \tag{6.18}$$

with

$$[B] = [B_M]^{-1} [B_J].$$ (6.19)

It should be noted that the present formulation provides for one level of smoothness beyond that actually required for the solution of the boundary condition equation. Since second derivatives exist for the spline function, simple point matching could have been used to solve the boundary condition equation. Alternatively, triangles could have been used as the expansion functions along with pulse testing.

Next the method of moments is used to solve the magnetic field integral equation (MFIE), which has been formulated above. Again the equation is tested using the pulse functions described earlier, resulting in the following matrix equation

$$\{I\} = [W] \{J\} + [Y] \{M\},$$ (6.20)

with the matrix and vector elements given by

$$I(i) = \oint_S P_i(t) \, \hat{a}_z \cdot \mathbf{H}_1^i(t) \, dt,$$ (6.21)

$$W(i,j) = -\oint_S P_i(t) \, \hat{a}_z \cdot \left[\nabla \times \oint_S S_j(t') \, \hat{a}_t(t') G(t,t') \, dt' \right] dt,$$ (6.22)

and

$$Y(i,j) = j\omega\epsilon \oint_S P_i(t) \left[\oint_S S_j(t') \, G(t,t') \, dt' \right] dt.$$ (6.23)

As always, careful attention must be paid to the singularities when evaluating the self terms in the matrices [44].

The final step is to substitute the relationship from the boundary condition into the above equation. This gives

$$\{J\} = [W^T]^{-1} \{I\},$$ (6.24)

with

$$[W^T] = [W] + [Y] [B].$$ (6.25)

It should be pointed out that the matrices $[Y]$ and $[W]$ depend only on the geometry of the scatterer, while the boundary condition matrix $[B]$ depends only on the coating parameters if the planar approximation is used, and on the coating parameters as well as the local radius of curvature if the higher order impedance boundary conditions that account for the curvature are used. When the scatterer is a perfect electric conductor (PEC), $[B] = 0$. If a Standard Impedance Boundary Condition (SIBC) is used, the matrix $[B]$ is a diagonal matrix. In all cases the evaluation of the $[B]$ matrix represented only a small fraction of the CPU time required for solution of the problem, with evaluation of the matrices $[Y]$ and $[W]$ dominating the total CPU time.

6.2   Exact Solution

A method of moments solution based on an exact formulation of the problem, which matches the tangential electric and magnetic fields across the dielectric to free space boundary and employs a CFIE solution on the inner conductor, is used to provide the "exact" solution to which the solutions based upon the approximate boundary conditions are compared. For this solution triangle functions are used to expand all three sets of currents, M and J on the free space dielectric boundary as well as J on the inner conductor, and they are also used as the testing functions. The details of the numerical implementation of the solution based on the exact formulation are omitted for the sake of brevity, but may be obtained from the literature [7].

6.3   Superquadric Cylinders

As was described earlier, conducting cylinders with a uniform lossy dielectric coating have been chosen as the scattering objects for this study. The external contour (dielectric-free space boundary) of these objects has been chosen to have the form of a superquadric cylinder. The superquadric cylinder is described through the elementary equation [45]

$$\left|\frac{y}{a}\right|^{\gamma} + \left|\frac{z}{b}\right|^{\gamma} = 1. \tag{6.26}$$

The case $\gamma = 2$ corresponds to an ellipse, while the case $\gamma \to \infty$ corresponds to a rectangle. It should be pointed out that since the dielectric to air boundary has been chosen to be superquadric and the dielectric layer is uniform, the internal conducting surface is not, in general, superquadric.

Some examples of superquadric cylinders are shown in Figure 6.5. Note that as $\gamma \to \infty$ the radius of curvature on each of the corners of the cylinder approaches zero. Since the approximate boundary conditions based on the planar approximation become invalid for small radii of curvature, the superquadric cylinder is ideal for studying this effect.

6.4   Examples

In this section computations of radar cross section of several dielectric-coated conducting cylinders are presented. Solutions based on a method of moments formulation that accounts for the dielectric exactly are compared to solutions based on the SIBC and the higher order impedance boundary conditions, which were derived earlier. Scatterers with dimensions in the range of two wavelengths are chosen as examples. In addition, the parameters of the coating are chosen to be in the region where the SIBC is known to fail. It is well known that for high values of the refractive index, high loss, and thin coatings, the SIBC is an excellent approximation, and hence these cases are not considered. These

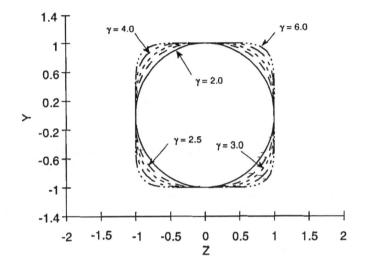

Figure 6.5: Examples of superquadric cylinders.

examples will demonstrate the improved accuracy of the higher order boundary conditions that include the effects of curvature relative to those based on a planar approximation.

6.4.1 Impedance Approximation

The first step in the solution of a scattering problem using the approximate boundary conditions is finding the coefficients c_0, c_1, and c_2, which appear in the ratio of polynomial approximation of the impedance. This may be accomplished by matching the exact impedance at three points, thus determining the three coefficients, or by using a least squares curve fitting procedure. The coating of interest for most of the examples of this section has the following parameters: $\epsilon_r = 4.0 - j0.5, \mu_r = 1.0$, and $d = 0.1\lambda_0$.

The real and imaginary parts of the exact impedance contained in Eq. (6.7), and the fitted impedance, Eq. (6.8) (dots), are shown for several radii of curvature in Figure 6.6. For example, the constants for the fitted impedance for the planar approximation are given by $c_0 = 113.62 + j555.7$ Ω, $c_1 = 0.18 + j3.36$ $\lambda^2\Omega$, and $c_2 = -0.01 + j0.002$ λ^2. When curvature is considered a slightly modified set of coefficients is determined. For example when the exterior radius, b, is equal to λ, a case of interest in the following section, we find that $c_0 = 89.6 + j496.0$ Ω, $c_1 = 0.11 + j3.3$ $\lambda^2\Omega$, and $c_2 = -0.01 + j0.002$ λ^2. For radii of curvature greater than five wavelengths the impedance is well approx-

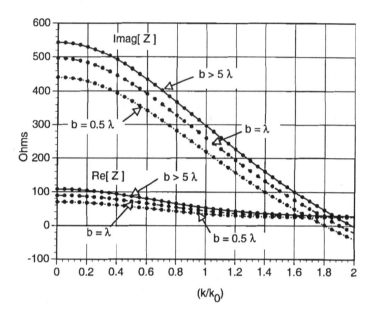

Figure 6.6: Exact and approximate (dots) impedance functions for several radii of curvature when $\epsilon_r = 4.0 - j0.5$, $\mu_r = 1.0$, and $d = 0.1\lambda_0$.

imated by that obtained from the planar problem. As can be seen from the figure, the fitted impedance function is an excellent approximation to the exact impedance in all cases. For example, for the planar case the maximum error is approximately 0.3% over the entire range of k_t. This figure may also be used to illustrate the inadequacy of the SIBC. The Standard Impedance Boundary Condition approximates the impedance by its value at $k_t = 0$. This is obviously a poor approximation for fields having any significant wavenumber in a direction parallel to the boundary.

6.4.2 Circular Cylinder: Bistatic RCS

In order to validate the method of moments computer codes, the case of a dielectric-coated circular cylinder was considered. This is a special case of the superquadric cylinder ($\gamma = 2.0$), for which exact series solutions may be obtained (Appendix E). The computed bistatic RCS for a coated circular cylinder with parameters $a = b = 1\lambda_0$, $d = 0.1\lambda_0$, $\epsilon_r = 4.0 - j0.5$, and $\mu_r = 1.0$ is shown in Figure 6.7. The backscatter direction is $\phi = 180$ degrees. Results for the exact formulation; two SIBC formulations, one including curvature, and one neglecting it; the formulation based upon the planar higher-order impedance boundary condition, HOIBC; as well as the formulation considering the HOIBC that includes the effects of curvature are presented in the figure. In all cases 15 basis functions per wavelength were used to represent the currents of interest

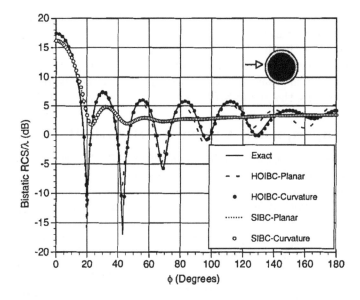

Figure 6.7: Bistatic RCS for a coated circular cylinder ($\gamma = 2$) when $a = b = \lambda_0$, $\epsilon_r = 4.0 - j0.5$, $\mu_r = 1.0$, and $d = 0.1\lambda_0$, with H polarization. The backscatter direction is 180 degrees.

in the method of moments solutions. The results from the method of moments solution were found to be in excellent agreement with the series solutions in all three cases, thus validating the computer codes.

As can be seen in the figure, the results using the planar HOIBC are in excellent agreement with the exact solution over most of the angular range, while both SIBC solutions are nearly identical to each other and give only the average behavior of the scattered field. The SIBC solutions do not give accurate predictions for the forward scattered field ($\phi = 0$ degrees), while the planar HOIBC solution result closely matches the exact solution. Both the SIBC solutions and the planar HOIBC solution are less accurate in the backscatter direction ($\phi = 180$ degrees) than in the forward scatter direction, for this example.

Additional computations have also been completed and indicate that, as expected, the accuracy of both the SIBC and the planar HOIBC improves as the size of the cylinder increases. It is also seen that the HOIBC that correctly accounts for the curvature gives results virtually identical to the exact solution. As was discussed in the previous section, for this particular case, since the radius of curvature is constant, the coefficients in the differential equation are constants, just as they were for the planar approximation. They are, however, not the same constants used in the planar case. This example illustrates that correcting the SIBC for curvature effects does not significantly increase its accuracy. As was discussed in Chapter 5, the primary weakness of the SIBC is its inability to predict the correct behavior of the coating for fields having significant wavenumbers parallel to the boundary. This fault is in no way remedied by the simple curvature correction. Since the SIBC that includes the effects of curvature produces results negligibly different from those of the planar SIBC for the examples considered in this chapter, the remaining results will consider only the planar SIBC in order to simplify the RCS plots.

6.4.3 Superquadric Cylinders: Bistatic RCS

The next three figures show results for the bistatic RCS for several superquadric dielectric-coated cylinders. Results for a cylinder with $a = b = 1\lambda_0$, $\gamma = 4.0$, and the same coating as that of the previous example are depicted in Figure 6.8. This cylinder, which is depicted in Figure 6.5, is illuminated by a plane wave,

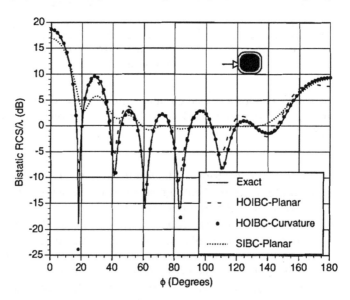

Figure 6.8: Bistatic RCS for a superquadric cylinder with $\gamma = 4$ when $a = b = \lambda_0$, $\epsilon_r = 4.0 - j0.5$, $\mu_r = 1.0$, and $d = 0.1\lambda_0$, with H polarization. Backscatter direction is 180 degrees.

and once again the backscatter direction is $\phi = 180$ degrees. As was the case for the circular cylinder example, the SIBC approximation gives only the average behavior in the sidelobe region, while the planar HOIBC gives highly accurate results. The relative levels of the sidelobes and the enhanced backscatter relative to the case of a dielectric-coated circle are predicted well. In this case the SIBC gives superior results only in a small angular range in the backscatter direction. Even with the presence of relatively sharp corners the planar HOIBC appears to be quite accurate. Once again the HOIBC that accounts for the curvature produces results virtually identical to the exact solution. In this case the coefficients in the differential equation are functions of position, due to the changes in the radius of curvature of the cylinder. The real and imaginary components of the coefficient c_0, as a function of position on the cylinder, are shown in Figure 6.9. Also plotted is the value of c_0 that is used in the planar

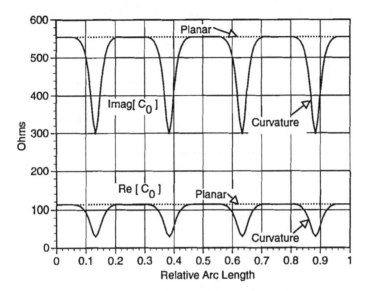

Figure 6.9: Real and imaginary parts of the coefficient c_0 as a function of position on the cylinder for the planar HOIBC (dashed) and the curvature HOIBC (solid).

approximation. The two coefficients are similar only on regions of the cylinder that are essentially flat. The remaining coefficients, c_1 and c_2, behave similarly.

A more severe test of the HOIBC is depicted in Figure 6.10, where $a = b = 1\lambda_0$ and $\gamma = 6.0$. In this case the cylinder is nearly square, as can be seen in Figure 6.5. Due to the small radii of curvature at the corners of the cylinder, the results for the planar HOIBC are more inaccurate than for the $\gamma = 4.0$ case. The planar HOIBC solution is still able to predict the positions of the pattern lobes and their relative heights. The backscatter is also accurately predicted. Once again the SIBC result is unacceptable. Even for this difficult example the agreement between the results using the higher-order impedance boundary conditions that consider curvature and the exact solution is excellent.

Since the corner of the scatterer has the smallest radius of curvature, and the higher-order impedance boundary conditions are expected to be the most inaccurate there, an example where the plane wave is incident on the corner is considered in Figure 6.11. The figure shows the computed bistatic RCS for the same superquadric cylinder as Figure 6.8, $\gamma = 4.0$, $a = b = 1\lambda_0$, and the

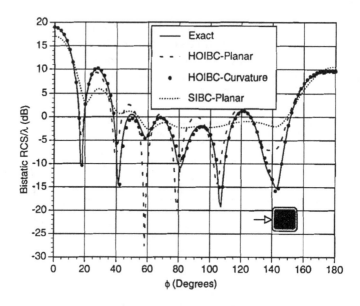

Figure 6.10: Bistatic RCS for a superquadric cylinder with $\gamma = 6$ when $a = b = \lambda_0$, $\epsilon_r = 4.0 - j0.5$, $\mu_r = 1.0$, and $d = 0.1\lambda_0$, with H polarization. Backscatter direction is 180 degrees.

same coating, but with an angle of incidence of 45 degrees. The agreement between the planar HOIBC and the exact solution is excellent, particularly in the forward scatter direction $\phi = 225$ degrees. All of the sidelobes are predicted quite well by the planar HOIBC solution, while the SIBC gives only the average behavior. The result in the backscatter direction, 45 degrees, is in error by approximately 2 dB for the planar higher-order impedance boundary condition, and substantially more for the SIBC solution. The HOIBC that considers curvature gives essentially the exact result in all directions, including the backscatter direction.

6.4.4   Superquadric Cylinders: Monostatic RCS

An example of monostatic RCS is given for a rectangular-like superquadric cylinder in Figure 6.12. For this example a cylinder with the same coating as the previous examples was used, $\gamma = 6.0$, $a = 1\lambda_0$, and $b = 1.5\lambda_0$. As was expected, when the plane wave is incident on the flat portions of the cylinder reasonably accurate results are obtained in all cases. However, both the SIBC and the planar higher-order impedance boundary condition formulations fail to give an accurate prediction of the backscattered field for incidence on portions of the scatterer where the radius of curvature is small. By including the effects of curvature essentially exact results are obtained for all angles of incidence. The

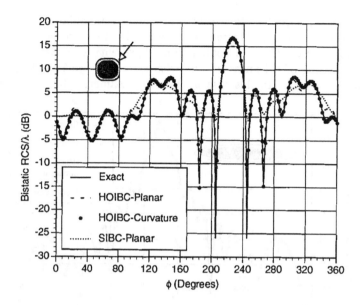

Figure 6.11: Bistatic RCS of a superquadric cylinder with $\gamma = 4$ when $a = b = \lambda_0$, $\epsilon_r = 4.0 - j0.5$, $\mu_r = 1.0$, and $d = 0.1\lambda_0$, with H polarization. Backscatter direction is 45 degrees.

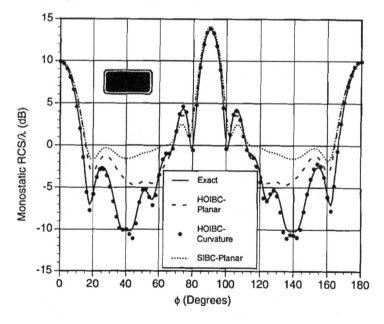

Figure 6.12: Monostatic RCS of a superquadric cylinder with $\gamma = 6$ when $a = 1.5\lambda_0$, $b = \lambda_0$, $\epsilon_r = 4.0 - j0.5$, $\mu_r = 1.0$, and $d = 0.1\lambda_0$, with H polarization.

accuracy of the results for bistatic RCS for these rectangular coated cylinders is similar to the results for the square cylinders discussed earlier.

The monostatic RCS of a coated thin ellipse with $a = 2\lambda_0$, $b = 0.25\lambda_0$, a uniform coating of thickness $0.05\lambda_0$, with $\epsilon_r = 4.0 - j0.5$, and $\mu_r = 1.0$ is shown in Figure 6.13. In this case all three approximate solutions are in close

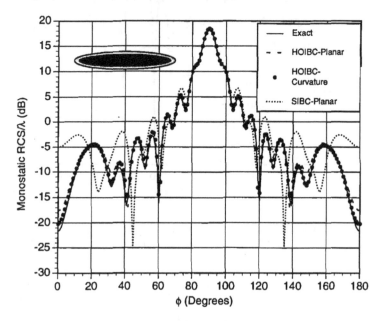

Figure 6.13: Monostatic RCS of a superquadric cylinder (ellipse) with $\gamma = 2$ when $a = 2.0\lambda_0$, $b = 0.25\lambda_0$, $\epsilon_r = 4.0 - j0.5$, $\mu_r = 1.0$, and $d = 0.05\lambda_0$, with H polarization.

agreement with the exact solution when the wave is incident on the broad face of the ellipse ($\phi = 90$ degrees). For angles of incidence more than 30 degrees in either direction from broadside incidence, the SIBC is no longer accurate. Both higher-order impedance boundary condition approximations are quite accurate for all angles of incidence. As expected, it is seen that for angles of incidence directly on the tip, including the effects of curvature improve the accuracy of the solution.

6.4.5   Superquadric Cylinder: Magnetic Coating

As a final example a superquadric cylinder, $\gamma = 4.0$, $a = b = 1\lambda_0$, with a magnetic coating, is considered. The parameters of the coating are $d = 0.1\lambda_0$, $\mu_r = 4.0 - j0.5$, and $\epsilon_r = 1.0$. The bistatic RCS is shown in Figure 6.14, and the backscatter direction is 180 degrees. For this high impedance case the planar higher-order impedance boundary condition and the SIBC both give relatively

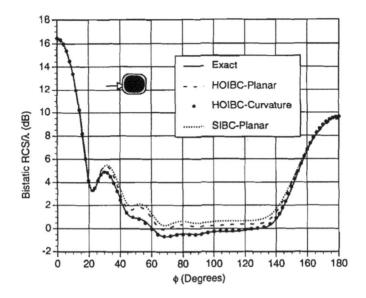

Figure 6.14: Bistatic RCS of a superquadric cylinder with $\gamma = 4$ when $a = b = \lambda_0$, $\epsilon_r = 1.0$, $\mu_r = 4.0 - j0.5$, and $d = 0.1\lambda_0$, with H polarization. Backscatter direction is 180 degrees.

accurate results for all angles, with both being most accurate in the backscatter and forward scatter regions. Including the curvature improves the results significantly. In this case the higher-order impedance boundary conditions give only a minor improvement in the results relative to those of the SIBC.

6.5   Conclusions

The examples presented indicate that the higher-order impedance boundary conditions obtained neglecting curvature give results for the bistatic cross section that are far superior to those of the SIBC. When the effects of curvature are included in the derivation of the higher-order impedance boundary conditions, the results are improved dramatically. Essentially perfect agreement between the results using these improved boundary conditions and the exact solution was obtained for several examples. In these cases the higher-order impedance boundary conditions provide high accuracy, while significantly reducing the number of unknowns required to model the problem. Excellent results have been obtained even for scatterers possessing corners with rather small radii of curvature, particularly when curvature is accounted for in the derivation of the approximate boundary condition.

Chapter 7

Scattering by Dielectric-Coated Bodies of Revolution

In this chapter we consider the specific case of scattering by coated conducting bodies of revolution (BOR). An exact formulation of this problem has been discussed in Ref. [7], where the required unknowns are the electric current on the inner conductor and equivalent electric and magnetic currents at the interface between the various layers making up the coating. Thus for an N-layer coating there are $2N + 1$ vector unknowns for an exact solution of the problem. For large bodies with multiple layers the number of unknowns can become significant, limiting the size of the object that may be considered.

In an effort to reduce the number of unknowns approximate boundary conditions may be used to model the behavior of the coating. In this case a relationship between the equivalent electric and magnetic currents on the outermost boundary of the body is deduced. This relationship along with the Electric Field Integral Equation (EFIE), Magnetic Field Integral Equation (MFIE), or the Combined Field Integral Equation (CFIE) is used to solve the scattering problem, using only two vector unknowns.

Implementation of the SIBC on bodies of revolution has been discussed by a number of authors [13-16]. The SIBC is a good approximation when the coating is thin with respect to a wavelength, has a high index of refraction, or has significant loss. Under these circumstances the electromagnetic field inside the coating propagates normal to the boundary, independent of the incident field, and therefore consideration of the coating's properties for normal incidence is sufficient. In practice, many coatings do not meet these criteria. In this chapter HOIBC are applied to the problem of scattering by three-dimensional coated bodies of revolution [46].

After a brief description of the problem geometry, and the Fourier series decomposition of the problem, a brief review of the exact formulation for scattering by coated bodies of revolution is presented. Next the higher order impedance boundary condition formulation of the problem is discussed in detail. First the HOIBC for planar coatings is described, and the SIBC is shown to be included in the HOIBC as a special case. Next the problem of transferring the planar HOIBC onto the doubly curved body of revolution (BOR) is considered, and the errors introduced during this process are discussed. The numerical solution of the resulting differential and integral equations is discussed, and the form of the system matrix is compared to that obtained from the exact and SIBC formulations of the problem. A number of examples comparing the exact, SIBC, and HOIBC solutions are used to demonstrate the effects of curvature and coating parameters on the accuracy of the HOIBC solution. Some general guidelines are given regarding the applicability of the HOIBC solution to a given problem.

7.1   Problem Geometry

The geometry for a coated conducting (BOR) is depicted in Figure 7.1. The nth layer of material is bounded by outer and inner surfaces, s_{n-1} and s_n respectively, which are generated by rotating the two parametric curves c_{n-1} and c_n around the z axis. On the outer boundary c_{n-1}, $\rho(t) = \rho_{n-1}(t)$ and $z(t) = z_{n-1}(t)$ with $0 \leq t \leq l_{n-1}$, and on the inner boundary c_n, $\rho(t) = \rho_n(t)$, and $z(t) = z_n(t)$ with $0 \leq t \leq l_n$, with l_n being the arc length of the nth parametric curve. The curve c_0 therefore represents the outer boundary of the BOR, and c_N is the boundary of the conducting core for an N layer coating. On each surface, unit vectors $\hat{t}$, $\hat{\phi}$, and $\hat{a}_n = \hat{\phi} \times \hat{t}$ are defined as shown in Figure 7.1.

Due to the rotational symmetry of the problem an efficient solution may be obtained by expanding the incident fields, the scattered fields, and their sources, in a Fourier series in the azimuthal, ϕ, direction. The overall problem then decouples and each of the terms in the series may be found independently.

7.2   Fourier Series Decomposition

The numerical implementation of any exact or HOIBC formulation begins with the piecewise linear approximation to each of the curves. Such an approximation

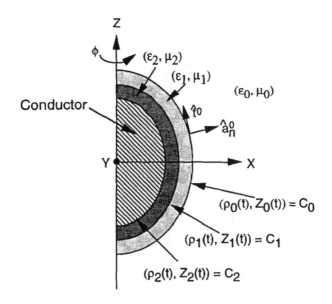

Figure 7.1: Geometry for a coated conducting body of revolution.

for the exterior contour C_0 is shown in Figure 7.1. Each curve is described by J_T linear segments, and $J_T + 1$ points as shown in Figure 7.2. The first and last points must lie on the axis of rotation in order to form a closed body. As was mentioned earlier, the unknown currents as well as the incident field are expanded into a Fourier series in ϕ, thus breaking up the large three-dimensional scattering problem into several smaller ones. The currents are expanded in terms of a Fourier series in ϕ, and triangle functions along the generating curve, as follows,

$$\mathbf{J}(t,\phi) = \sum_{n=0}^{N} \sum_{j=1}^{J_T} \left(a_n^t(j)\hat{t} - a_n^\phi(j)\hat{\phi} \right) \frac{T_j(t)}{\rho(t)} e^{jn\phi} \tag{7.1}$$

and

$$\mathbf{M}(t,\phi) = \eta_0 \sum_{n=0}^{N} \sum_{j=1}^{J_T} \left(b_n^t(j)\hat{t} - b_n^\phi(j)\hat{\phi} \right) \frac{T_j(t)}{\rho(t)} e^{jn\phi}. \tag{7.2}$$

Here η_0 is the free space wave impedance and $T_k(t)$ is the kth triangle function, which spans four segments on the generating curve, as shown in Figure 7.2. The coefficients a_n^t, a_n^ϕ, b_n^t, and b_n^ϕ are the unknown quantities to be solved for. In general at least ten triangle functions are required per wavelength in each of the media in contact with the BOR generating curve, and therefore the curve must be divided up into at least twenty segments per wavelength, since there are approximately twice as many segments as triangles. It should be noted that

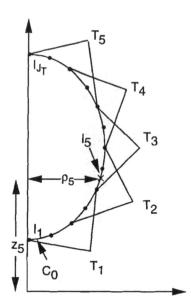

Figure 7.2: Piecewise linear approximation to the BOR generating curve, and triangle basis functions.

the quantities ρJ and ρM are expanded in terms of triangle functions, rather than the currents themselves. This implies that no half triangles are required near the poles of the BOR since the currents are finite there and ρ vanishes. The number of terms required in the Fourier series depends on the size of the body, and choosing $N \approx k_0 \rho_{max}$, where k_0 is the free space wavenumber, is typically sufficient. The next section will discuss the exact formulation of the problem.

7.3   Exact Solution

In this section we describe the formulation of the exact solution for the special case of a single layer coating. An extension to multilayer coatings is straightforward. A detailed description is given in Ref. [7]. The problem is solved by formulating equivalent problems for the external and internal regions and matching the tangential fields across the common boundary. For the external problem, depicted in Figure 7.3, we may replace the BOR by an equivalent set of electric, J_0^+, and magnetic, M_0^+, surface currents that radiate the same scattered field in the exterior region as the BOR. These currents reside on the S_0 surface, and the (+) superscript is used to signify that they along with the incident fields produce the true total fields in the exterior region and zero total fields in the interior region. The volume inside the BOR may be replaced with

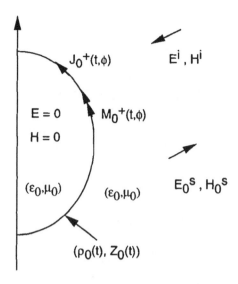

Figure 7.3: Equivalent problem for the exterior region.

free space since there are no fields there, and thus,

$$E_0 = E^i + E_0^s \left(M_0^+, J_0^+\right) \tag{7.3}$$

and

$$H_0 = H^i + H_0^s \left(M_0^+, J_0^+\right), \tag{7.4}$$

where the scattered fields are obtained from the equivalent sources using the appropriate free space Green's functions.

The exact internal problem for the one-layer case is depicted in Figure 7.4. The equivalent currents J_0^-, M_0^-, and J_1^+ generate the true field inside the layer and zero fields elsewhere. The current M_1^+ is absent due to the presence of the conducting core. The regions outside the layer are now filled with a material having the same electrical properties as the layer. In this case the incident field is zero, and the total field is the scattered field generated by the sources,

$$E_1 = E_1^s \left(M_0^-, J_0^-\right) + E_1^s \left(J_1^+\right) \tag{7.5}$$

and

$$H_1 = H_1^s \left(M_0^-, J_0^-\right) + H_1^s \left(J_1^+\right), \tag{7.6}$$

where the scattered fields are now computed using the appropriate Green's function for the coating. Since the tangential electric and magnetic fields must be continuous across the boundary , $J_0^- = -J_0^+$ and $M_0^- = -M_0^+$. The interior and exterior field representations must also match across the common boundary,

$$\hat{a}_n^0 \times E_0\big|_{S_0^+} = \hat{a}_n^0 \times E_1\big|_{S_0^-} \tag{7.7}$$

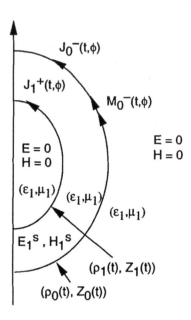

Figure 7.4: Exact equivalent interior problem for the one-layer case.

and

$$\hat{a}_n^0 \times \mathbf{H}_0\big|_{S_0^+} = \hat{a}_n^0 \times \mathbf{H}_1\big|_{S_0^-}, \tag{7.8}$$

where S_0^+, and S_0^- indicate that the fields are to be evaluated just outside $(+)$ or just inside $(-)$ the S_0 surface. An additional equation is obtained by satisfying either the EFIE,

$$\hat{a}_n^1 \times \mathbf{E}_1\big|_{S_1^-} = 0, \tag{7.9}$$

the MFIE,

$$\hat{a}_n^1 \times \mathbf{H}_1\big|_{S_1^-} = 0, \tag{7.10}$$

or a linear combination of these, the CFIE, just inside the conducting core. Use of the CFIE and the two previous equations, which match the interior and exterior representations of the tangential electric and magnetic fields across the common boundary, avoids any difficulties with cavity resonance frequencies [7].

Upon substituting the expressions for the unknown currents, and testing the equations using Galerkin's method, the final form of the matrix equation for the exact solution is given by,

$$
\begin{bmatrix}
[P_J^{tt}] & [P_J^{t\phi}] & [P_M^{tt}] & [P_M^{t\phi}] & [S_J^{tt}] & [S_J^{t\phi}] \\
[P_J^{\phi t}] & [P_J^{\phi\phi}] & [P_M^{\phi t}] & [P_M^{\phi\phi}] & [S_J^{\phi t}] & [S_J^{\phi\phi}] \\
[Q_J^{tt}] & [Q_J^{t\phi}] & [Q_M^{tt}] & [Q_M^{t\phi}] & [T_J^{tt}] & [T_J^{t\phi}] \\
[Q_J^{\phi t}] & [Q_J^{\phi\phi}] & [Q_M^{\phi t}] & [Q_M^{\phi\phi}] & [T_J^{\phi t}] & [T_J^{\phi\phi}] \\
[R_J^{tt}] & [R_J^{t\phi}] & [R_M^{tt}] & [R_M^{t\phi}] & [U_J^{tt}] & [U_J^{t\phi}] \\
[R_J^{\phi t}] & [R_J^{\phi\phi}] & [R_M^{\phi t}] & [R_M^{\phi\phi}] & [U_J^{\phi t}] & [U_J^{\phi\phi}]
\end{bmatrix}
\begin{Bmatrix}
\{a_0^t\} \\
\{a_0^\phi\} \\
\{b_0^t\} \\
\{b_0^\phi\} \\
\{a_i^t\} \\
\{a_i^\phi\}
\end{Bmatrix}
=
\begin{Bmatrix}
\{E^t\} \\
\{E^\phi\} \\
\{H^t\} \\
\{H^\phi\} \\
\{0\} \\
\{0\}
\end{Bmatrix}
$$

$$(7.11)$$

The entries in the various submatrices are discussed in detail in Ref. [7]. As seen above, the exact solution requires the determination of three unknown vector quantities. For a thin coating the number of triangles required on the inner conductor is approximately equal to that required on the outer boundary of the BOR, resulting in approximately $6J_T$ unknowns for an exact solution of the problem. It should also be noted that the above system matrix is dense, with nearly all of the entries being obtained through numerical integration. In an effort to remove some of these undesirable features we pursue the implementation of the HOIBC solution in the next section.

7.4   HOIBC Solution

7.4.1   Exterior Region

The Higher Order Impedance Boundary Condition formulation begins by considering the identical problem as that for the exact formulation depicted in Figure 7.3. A Combined Field Integral Equation is formed by taking a linear combination of the EFIE and the MFIE,

$$
\hat{a}_n^0 \times \mathbf{E}_0\big|_{S_0^-} + \frac{\alpha}{1-\alpha}\left(\hat{a}_n^0 \times \hat{a}_n^0 \times \mathbf{H}_0\big|_{S_0^-}\right) = 0. \tag{7.12}
$$

When $\alpha = 0$ the CFIE becomes the EFIE, and as $\alpha \to 1$ the CFIE tends to the MFIE. The CFIE is the integral equation of choice for the exterior problem, since it contains sufficient information to eliminate the problem of spurious resonances associated with the cavity formed by the exterior contour C_0 [13]. The fields at a point $\mathbf{r}$ may be computed from a set of sources on the surface S_0, with position vector $\mathbf{r}'$, using two operators L and K, which contain the free space Green's function [7],

$$
\mathbf{E}_0^s(\mathbf{r}) = K\left[\mathbf{M}_0^+(\mathbf{r}')\right] - L\left(\mathbf{J}_0^+(\mathbf{r}')\right) \tag{7.13}
$$

and

$$
\mathbf{H}_0^s(\mathbf{r}) = -K\left[\mathbf{J}_0^+(\mathbf{r}')\right] - \frac{1}{\eta_0^2}L\left[\mathbf{M}_0^+(\mathbf{r}')\right], \tag{7.14}
$$

where

$$K\left[\mathbf{X}(\mathbf{r}')\right] = \int_{S_0} \mathbf{X}(\mathbf{r}') \times \nabla G_0(\mathbf{r}, \mathbf{r}') ds', \tag{7.15}$$

$$L\left[\mathbf{X}(\mathbf{r}')\right] = -j\omega\mu_0 \int_{S_0} \left(\mathbf{X}(\mathbf{r}') + \frac{1}{\omega^2 \mu_0 \epsilon_0} \nabla\left(\nabla' \cdot \mathbf{X}(\mathbf{r}')\right)\right) G_0(\mathbf{r}, \mathbf{r}') ds', \tag{7.16}$$

and the free space Green's function is given by

$$G_0(\mathbf{r}, \mathbf{r}') = \frac{e^{-jk_0|\mathbf{r}-\mathbf{r}'|}}{4\pi|\mathbf{r} - \mathbf{r}'|}. \tag{7.17}$$

As indicated above, the integral is carried out over the entire exterior surface of the BOR. When the observation point $\mathbf{r}$ and the source point $\mathbf{r}'$ coalesce the integrals must be interpreted in the principal value sense.

Upon substituting the previous expansions for the currents in Eqs. (7.1) and (7.2) into Eq. (7.12), multiplying by each of the weighting functions,

$$\mathbf{W}_n^t(i) = \frac{T_i(t)}{\rho(t)} e^{-jn\phi}\hat{t} \tag{7.18}$$

and

$$\mathbf{W}_n^\phi(i) = \frac{T_i(t)}{\rho(t)} e^{-jn\phi}\hat{\phi}, \tag{7.19}$$

and integrating over the surface S_0, $2J_T$ equations are obtained.

These equations may be written in matrix form as

$$\{V^t\} = [Z_J^{tt}]\{a^t\} + [Z_J^{t\phi}]\{a^\phi\} + [Z_M^{tt}]\{b^t\} + [Z_M^{t\phi}]\{b^\phi\} \tag{7.20}$$

and

$$\{V^\phi\} = [Z_J^{\phi t}]\{a^t\} + [Z_J^{\phi\phi}]\{a^\phi\} + [Z_M^{\phi t}]\{b^t\} + [Z_M^{\phi\phi}]\{b^\phi\}. \tag{7.21}$$

Expressions for the individual matrix elements are included in Appendix F. As with the exact solution the matrices are dense with the majority of the entries requiring numerical integration. Additional $2J_T$ equations are required to solve for the complete set of unknowns. These equations are provided by the higher order impedance boundary condition discussed in the next section.

7.4.2   Interior Region

As was mentioned previously, higher order impedance boundary conditions can be used to relate the tangential electric and magnetic fields on the surface of a coated body. The HOIBC is represented as a pair of differential equations that relate the tangential field components, and hence the equivalent currents, $\mathbf{J}_0^+$ and $\mathbf{M}_0^+$.

The procedure of Chapter 2 may be used to derive the following pair of differential equations, relating the tangential components of the field on the exterior of an infinite dielectric coated ground plane,

$$\left(1 - c_2 \frac{\partial^2}{\partial x^2} - c_1 \frac{\partial^2}{\partial y^2}\right) E_x(x,y) + (c_1 - c_2) \frac{\partial^2}{\partial x \partial y} E_y(x,y)$$

$$= (c_4 - c_5) \frac{\partial^2}{\partial x \partial y} H_x(x,y) + \left(-c_3 + c_5 \frac{\partial^2}{\partial x^2} + c_4 \frac{\partial^2}{\partial y^2}\right) H_y(x,y) \qquad (7.22)$$

and

$$(c_1 - c_2) \frac{\partial^2}{\partial x \partial y} E_x(x,y) + \left(1 - c_1 \frac{\partial^2}{\partial x^2} - c_2 \frac{\partial^2}{\partial y^2}\right) E_y(x,y)$$

$$= \left(c_3 - c_4 \frac{\partial^2}{\partial x^2} + c_5 \frac{\partial^2}{\partial y^2}\right) H_x(x,y) - (c_4 - c_5) \frac{\partial^2}{\partial x \partial y} H_y(x,y) \qquad (7.23)$$

The above spatial domain boundary condition is now assumed to apply on a local basis, and thus the constants $c_1 \ldots c_5$ are functions of position, depending on the local electromagnetic properties of the coating material as well as its thickness d. The above differential equations reduce to simple linear equations that express the SIBC when all of the coefficients except c_3 are set equal to zero.

For multilayer coatings the impedances Z_{xy} and Z_{yx} must be replaced by their multilayer counterparts, which may be obtained from simple transmission line methods. All other parts of the HOIBC derivation remain the same. As was shown in Chapters 2 and 3, for coatings containing other materials such as chiral material, the HOIBC is slightly more complicated, but it may still be expressed in terms of a pair of second order differential equations.

The next task is to transfer these differential equations onto the BOR. Since the body is rotationally symmetric the coefficients in the resulting differential equations are independent of ϕ, and it is sufficient to consider the contour along the BOR given by $\phi = 0$. We begin this process by placing a local rectangular coordinate system (x', y', z') on the BOR with x' pointed in the ϕ direction, y' pointed in the t direction, and z' pointed in the direction of the local normal. The lines of constant x' and y' are projected onto the BOR as shown in Figure 7.5. This produces a set of curves on the surface of the BOR that in general intersect at right angles only at one point. The projections of constant y' will be referred to as type 1 curves, and projections of constant x' as type 2 curves. The differential Eqs. (7.22) and (7.23) are now transformed onto the curved coordinate system by using $\partial/\partial x \rightarrow \partial/\partial l_1$, $\partial/\partial y \rightarrow \partial/\partial l_2$, $V_x \rightarrow V_1$, and $V_y \rightarrow V_2$, where l_1 is the arc length along the type 1 curve, l_2 is the arc length along the type two curve, and V_1, V_2 are the vector components pointing in the directions of the type 1 and 2 curves. The appropriate differential equations become

$$\left(1 - c_2 \frac{\partial^2}{\partial l_1^2} - c_1 \frac{\partial^2}{\partial l_2^2}\right) E_1(l_1, l_2) + (c_1 - c_2) \frac{\partial^2}{\partial l_1 \partial l_2} E_2(l_1, l_2)$$

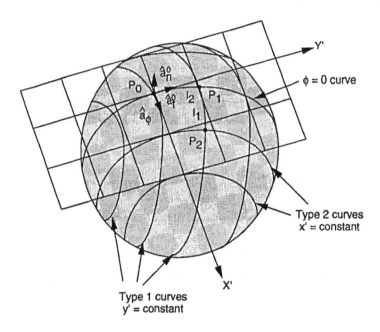

Figure 7.5: Transforming the planar higher order impedance boundary condition (HOIBC) onto the curved BOR.

$$= (c_4 - c_5) \frac{\partial^2}{\partial l_1 \partial l_2} H_1(l_1, l_2) + \left(-c_3 + c_5 \frac{\partial^2}{\partial l_1^2} + c_4 \frac{\partial^2}{\partial l_2^2} \right) H_2(l_1, l_2) \qquad (7.24)$$

and

$$(c_1 - c_2) \frac{\partial^2}{\partial l_1 \partial l_2} E_1(l_1, l_2) + \left(1 - c_1 \frac{\partial^2}{\partial l_1^2} - c_2 \frac{\partial^2}{\partial l_2^2} \right) E_2(l_1, l_2)$$
$$= \left(c_3 - c_4 \frac{\partial^2}{\partial l_1^2} + c_5 \frac{\partial^2}{\partial l_2^2} \right) H_1(l_1, l_2) - (c_4 - c_5) \frac{\partial^2}{\partial l_1 \partial l_2} H_2(l_1, l_2) \qquad (7.25)$$

All of the derivatives in the above expressions are to be evaluated at the tangent point, henceforth denoted as P_0, where $l_1 = l_2 = 0$.

The above transformation is exact only if the principal radii of curvature at the point of interest are infinite. Since curvature effects were not included in the determination of the coefficients, additional errors will occur over and above those due to the inherent approximate nature of the planar HOIBC. These errors will be proportional to the amount of curvature present at each point on the BOR. Unlike the two-dimensional examples considered in Chapter 6, the three-dimensional body of revolution possesses finite radii of curvature in two directions and the right circular cylinder is no longer the appropriate canonical problem for determining the higher order impedance boundary conditions that include curvature effects. Extension of the results of Chapter 4 to the three-dimensional case requires examination of a new canonical problem such as scattering by a coated spheroid. This problem is considerably more difficult than the cylinder problem, and construction of the appropriate HOIBC for doubly curved surfaces has been left as an area for future research.

Since the field and current quantities are defined in terms of the t, ϕ coordinate system of the BOR, the above equations must now be transformed into these coordinates. In order to transform the boundary condition equations the unit vectors in the l_1 and l_2 directions must be found, and the differential operators must be transformed. The details of these computations are included in Appendix G. The resulting unit vectors are found to be

$$\hat{a}_1(t,\phi) = \left(\left(\frac{\partial \rho}{\partial t}\bigg|_{P_0} \frac{\partial \rho}{\partial t} \cos\phi + \frac{\partial z}{\partial t}\bigg|_{P_0} \frac{\partial z}{\partial t} \right) \hat{\phi} + \frac{\partial \rho}{\partial t}\bigg|_{P_0} \sin\phi\hat{t} \right) / D_1(t,\phi) \quad (7.26)$$

with

$$D_1(t,\phi) = \sqrt{ \left(\frac{\partial \rho}{\partial t}\bigg|_{P_0} \frac{\partial \rho}{\partial t} \cos\phi + \frac{\partial z}{\partial t}\bigg|_{P_0} \frac{\partial z}{\partial t} \right)^2 + \left(\frac{\partial \rho}{\partial t}\bigg|_{P_0} \sin\phi \right)^2 }, \quad (7.27)$$

and

$$\hat{a}_2(t,\phi) = \left(\cos\phi\hat{t} - \frac{\partial \rho}{\partial t} \sin\phi\hat{\phi} \right) / D_2(t,\phi) \quad (7.28)$$

with

$$D_2(t,\phi) = \sqrt{ \cos^2\phi + \left(\frac{\partial \rho}{\partial t} \right)^2 \sin^2\phi }. \quad (7.29)$$

The differential operators transform as

$$\frac{\partial}{\partial l_1}\bigg|_{P_0} \rightarrow \frac{1}{\rho_0} \frac{\partial}{\partial \phi} \quad (7.30)$$

$$\frac{\partial^2}{\partial l_1^2}\bigg|_{P_0} \rightarrow \frac{1}{\rho_0^2} \frac{\partial^2}{\partial \phi^2} + \frac{1}{\rho_0} \frac{d\rho}{dt}\bigg|_{P_0} \frac{\partial}{\partial t} \quad (7.31)$$

$$\frac{\partial}{\partial l_2}\bigg|_{P_0} \rightarrow \frac{\partial}{\partial t} \quad (7.32)$$

$$\frac{\partial^2}{\partial l_2^2}\bigg|_{P_0} \rightarrow \frac{\partial^2}{\partial t^2} \quad (7.33)$$

and

$$\frac{\partial^2}{\partial l_1 \partial l_2}\bigg|_{P_0} \rightarrow \frac{1}{\rho_0} \frac{\partial^2}{\partial \phi \partial t} - \frac{1}{\rho_0^2} \frac{d\rho}{dt}\bigg|_{P_0} \frac{\partial}{\partial \phi}. \quad (7.34)$$

Upon substitution of these expressions into Eqs. (7.24) and (7.25) the HOIBC equations may now be expressed in the t, ϕ coordinate system. These boundary condition equations are now solved using the method of moments.

We begin by recognizing that $E_\phi = M_t$, $E_t = -M_\phi$, $H_\phi = -J_t$, and $H_t = J_\phi$. The expansions of Eqs. (7.1), and (7.2) are then substituted into the HOIBC equation and tested using a testing function $T_i(t)\rho(t)$. The result may be written in matrix form as

$$[P_1]\{b^t\} + [P_2]\{b^\phi\} + [P_5]\{a^t\} + [P_6]\{a^\phi\} = 0 \qquad (7.35)$$

and

$$[P_3]\{b^t\} + [P_4]\{b^\phi\} + [P_7]\{a^t\} + [P_8]\{a^\phi\} = 0 \qquad (7.36)$$

Since a given triangle overlaps only itself and its nearest neighbors, the matrices $P_1 - P_8$ are tridiagonal. Their matrix entries are given in detail in Appendix H.

Having completed the solution of the boundary condition equations we may assemble the complete method of moments (MOM) matrix,

$$
\begin{bmatrix}
[Z_J^{tt}] & [Z_J^{t\phi}] & [Z_M^{tt}] & [Z_M^{t\phi}] \\
[Z_J^{\phi t}] & [Z_J^{\phi\phi}] & [Z_M^{\phi t}] & [Z_M^{\phi\phi}] \\
[P_5] & [P_6] & [P_1] & [P_2] \\
[P_7] & [P_8] & [P_3] & [P_4]
\end{bmatrix}
\left\{
\begin{array}{c}
\{b^t\} \\
\{b^\phi\} \\
\{a^t\} \\
\{a^\phi\}
\end{array}
\right\}
=
\left\{
\begin{array}{c}
\{V^t\} \\
\{V^\phi\} \\
\{0\} \\
\{0\}
\end{array}
\right\}
\qquad (7.37)
$$

The bottom half of this matrix represents the HOIBC part of the solution. It represents only a slight modification of that obtained using the SIBC. In the case of the SIBC $[P_1] = [P_4] = [I]$, $[P_2] = [P_3] = [P_5] = [P_8] = [0]$, and $[P_6] = -[P_7] = [\eta]$, where $[I]$ is the identity matrix, $[0]$ is the zero matrix, and $[\eta]$ is a diagonal matrix. We shall see in the results section that the additional terms contributed to the matrix by the HOIBC have a significant effect on the accuracy of the solution.

The above system matrix should be contrasted with that of the exact formulation, Eq. (7.11). The total number of nonzero entries in the HOIBC system matrix is approximately $8J_T^2 + 24J_T$, whereas the exact system matrix has $36J_T^2$ nonzero entries for a single layer. For a number of thin layers the size of the HOIBC matrix is unchanged, whereas the exact system matrix size increases rapidly. The next section will present some results from the exact, HOIBC, and SIBC formulations for some representative coated bodies of revolution.

7.5   Examples

In this section several examples will be given to demonstrate the accuracy of the higher order impedance boundary conditions under a variety of circumstances. The examples are chosen to include a number of different BOR shapes, coating thicknesses, relative permitivity, and relative permeability.

7.5.1   Coated Sphere: Validation

As a first example we consider the case of a coated conducting sphere with a conductor radius of $1.5\lambda_0$ and coating thickness of $0.075\lambda_0$, with $\epsilon_r = 5.0$ and $\mu_r = 1.0$. Figure 7.6 and Figure 7.7 show the $\theta\theta$ and $\phi\phi$ components of the bistatic RCS for a plane wave incident from $\theta = 0$. Four solutions

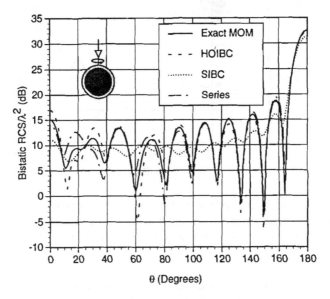

Figure 7.6: $\theta\theta$ component of the bistatic RCS for a coated conducting sphere with a conductor radius of $1.5\lambda_0$ and coating thickness $0.075\lambda_0$ with $\epsilon_r = 5.0$ and $\mu_r = 1.0$. $\theta = 0$ degrees corresponds to the backscatter direction.

are included: the exact series solution, a method of moments solution based on an exact formulation [7], the HOIBC solution, and a Standard Impedance Boundary Condition (SIBC) solution. The same approximate geometry was used to model the sphere in the MOM solutions where 160 facets were used to approximate the sphere. The slight differences between the series and exact MOM formulations can be attributed to the approximate nature of the faceted approximation to the sphere. The figures clearly show the increased accuracy of the HOIBC solution relative to the SIBC solution. The SIBC gives only the average behavior of the scattered field, while the HOIBC accurately predicts the sidelobe behavior. This behavior is quite similar to that noted in the previous chapter for two-dimensional coated bodies. It may also be noted from the figures that the maximum error in the scattered field is in the backscatter direction. This is also consistent with the two-dimensional results. In the two-dimensional case the backscatter error was shown to be reduced when curvature was included in the boundary condition. The remaining errors in Figure 7.6 and Figure 7.7 can be attributed to the fact that for a coated BOR both principal

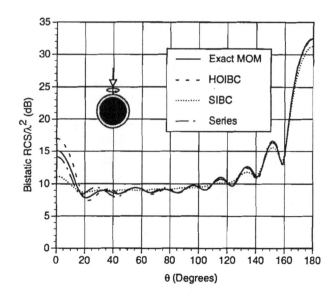

Figure 7.7: $\phi\phi$ component of the bistatic RCS for a coated conducting sphere with a conductor radius of $1.5\lambda_0$ and coating thickness $0.075\lambda_0$ with $\epsilon_r = 5.0$ and $\mu_r = 1.0$. $\theta = 0$ degrees corresponds to the backscatter direction.

radii of curvature are finite, whereas the present formulation of the HOIBC neglects curvature entirely. Additional computations have verified that the accuracy of the planar HOIBC improves as the radius of the sphere increases for a given set of coating parameters. Figure 7.8 shows the $\theta\theta$ bistatic RCS, for incidence at $\theta = 90$ degrees. It is included to illustrate that the HOIBC results for $\theta = 90$ degree incidence are identical to those for axial incidence. For the case of 90 degree incidence 12 Fourier components have been used in the computation. This illustrates the accuracy of the HOIBC for the higher order Fourier components.

7.5.2   Superquadric Cylinders: Curvature Effects

The next three examples will be used to illustrate the effects of curvature on the accuracy of the HOIBC solution. The coating under consideration is the same as in the previous examples, while the shape of the inner conductor portion of the BOR is given in Figure 7.9. Case (a) is a coated conducting cylinder of height λ_0 and radius λ_0. Cases (b) and (c) are bodies of revolution that will be denoted as superquadric bodies of revolution. They are formed by rotating superquadric curve about the z axis. The superquadric BOR are determined through the equation

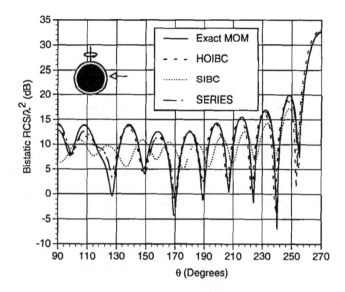

Figure 7.8: $\theta\theta$ component of the bistatic RCS for a coated conducting sphere with a conductor radius of $1.5\lambda_0$, coating thickness $0.075\lambda_0$ with $\epsilon_r = 5.0$ and $\mu_r = 1.0$. $\theta = 90$ degrees corresponds to the backscatter direction.

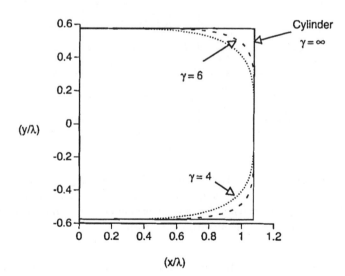

Figure 7.9: Geometry for several bodies of revolution generated from superquadric cylinders.

$$\left|\frac{\rho}{a}\right|^{\gamma} + \left|\frac{z}{b}\right|^{\gamma} = 1, \qquad (7.38)$$

which is closely related to that used in Chapter 6, Eq. (6.26). The parameter γ determines the shape of the resulting BOR. For $\gamma = 2$ we obtain spheroids, while as $\gamma \rightarrow \infty$ we obtain cylinders. Case (b) corresponds to $a = \lambda_0$, $b = 0.5$ λ_0, and $\gamma = 6$. Case (c) uses the same values for a and b with $\gamma = 4$. As can be seen in the figure, lowering the value of γ tends to ease the corner on the cylinder. Since the HOIBC has been based on a planar approximation we expect its accuracy to improve as the sharp corner is eased. The results of Figure 7.10, Figure 7.11, and Figure 7.12 illustrate that this is indeed the case. Here the $\theta\theta$ component of the bistatic RCS is plotted for axial incidence $\theta = 0$ degrees. The HOIBC, the SIBC and an exact MOM formulation are plotted.

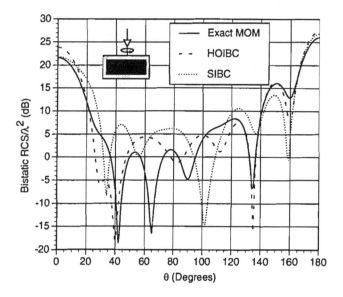

Figure 7.10: $\theta\theta$ component of the bistatic RCS for a coated conducting cylinder ($\gamma \rightarrow \infty$), with radius $= 1.0\lambda_0$, height $= \lambda_0$, and coating thickness $0.075\lambda_0$, with $\epsilon_r = 5$ and $\mu_r = 1.0$, illustrating the detrimental effects of a sharp corner. $\theta = 0$ degrees corresponds to the backscatter direction.

In all three cases 96 facets were used to approximate the BOR. The agreements for the $\phi\phi$ component are similar, generally being better than those for the $\theta\theta$ component for both the SIBC and HOIBC. For all three cases the results based on the SIBC are poor, while the results from the HOIBC improve with decreasing values of γ, as expected.

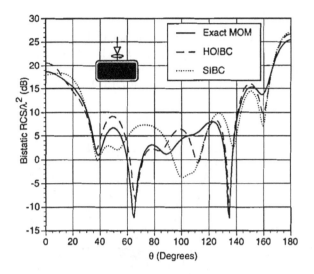

Figure 7.11: $\theta\theta$ component of the bistatic RCS for a coated conducting super-quadric BOR with $a = 1.0\lambda_0$, $b = 0.5\lambda_0$, $\gamma = 6.0$, and coating thickness $0.075\lambda_0$, with $\epsilon_r = 5$ and $\mu_r = 1.0$. $\theta = 0$ degrees corresponds to the backscatter direction.

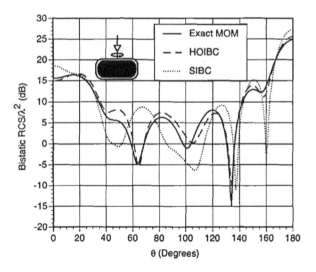

Figure 7.12: $\theta\theta$ component of the bistatic RCS for a coated conducting super-quadric BOR with $a = 1.0\lambda_0$, $b = 0.5\lambda_0$, $\gamma = 4.0$, and coating thickness $0.075\lambda_0$, with $\epsilon_r = 5$ and $\mu_r = 1.0$. $\theta = 0$ degrees corresponds to the backscatter direction.

7.5.3   Spheroid: Monostatic RCS

Since the HOIBC is based on the concept of "local" behavior we would expect its accuracy to be increased for lossy coatings. This is illustrated in Figure 7.13 and Figure 7.14, which present results for a coated oblate spheroid with semi-minor axis $0.5\lambda_0$ and semi-major axis λ_0. Figure 7.13 shows the $\theta\theta$ monostatic RCS

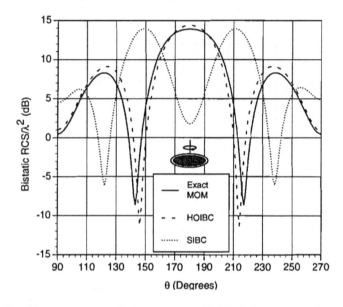

Figure 7.13: $\theta\theta$ component of the monostatic RCS for a coated conducting spheroid with a minor axis of $1.0\lambda_0$, major axis of $2.0\lambda_0$, and coating thickness $0.075\lambda_0$, with $\epsilon_r = 5.0$ and $\mu_r = 1.0$.

for a lossless coating, $d=0.075\lambda_0$, $\epsilon_r = 5.0$, and $\mu_r = 1.0$, whereas Figure 7.14 shows the RCS when $d=0.075\lambda_0$, $\epsilon_r = 4 - j0.5$, and $\mu_r = 1.0$. In both cases the HOIBC results are found to be superior to those of the SIBC. For the lossy coating excellent results are obtained for all angles of incidence. As was discussed earlier the finite radii of curvature on the spheroid contribute to the inaccuracy of the HOIBC results for both cases. In computing the results for the spheroid 96 facets were used to approximate the BOR, and 8 Fourier components were used in the solution.

7.5.4   Superquadric Cylinder: Magnetic Coating

As a final example a magnetic coating is considered. The inner conductor is chosen to be the same as in the previous section where $a = 1.0\lambda_0$, $b = 0.5\lambda_0$, and $\gamma = 4.0$, with coating parameters $d=0.06\lambda_0$, $\epsilon_r = 1.0$, and $\mu_r = 3.0$. Results for the $\theta\theta$ bistatic RCS, for incidence from $\theta = 0$ degrees, is shown in Figure 7.15. Once again the HOIBC results are seen to be quite superior to the SIBC results.

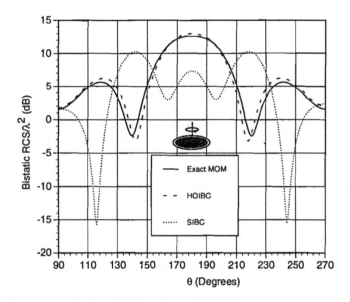

Figure 7.14: $\theta\theta$ component of the monostatic RCS for a coated conducting spheroid with a minor axis of $1.0\lambda_0$, major axis of $2.0\lambda_0$, and coating thickness $0.075\lambda_0$, with $\epsilon_r = 4.0 - j0.5$ and $\mu_r = 1.0$.

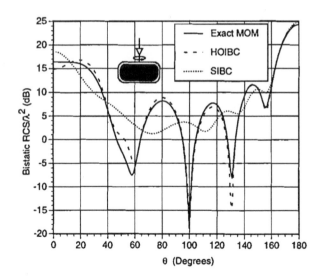

Figure 7.15: $\theta\theta$ component of the bistatic RCS for a coated conducting super-quadric BOR with $a = 1.0\lambda_0$, $b = 0.5\lambda_0$, $\gamma = 4.0$, and coating thickness $0.06\lambda_0$, with $\epsilon_r = 1.0$ and $\mu_r = 3.0$. $\theta = 0$ degrees corresponds to the backscatter direction.

7.6 Conclusions

A higher order impedance boundary condition solution for scattering by dielectric coated bodies of revolution has been presented. For the exterior problem, the solution employs the same expansion functions and testing procedure as commonly has been used in the past to solve the BOR scattering problem. The exact interior problem is eliminated through the use of the HOIBC, resulting in a significant reduction in the number of unknowns and computation time. Particular attention has been paid to the problem of transferring the planar HOIBC onto the doubly curved body of revolution. When compared to the results based on an exact formulation, the HOIBC results were found to be quite superior to those based on the SIBC. Since the curvature of the body of revolution was not accounted for in the HOIBC formulation, errors similar to those described when planar HOIBC were applied to two-dimensional coated circular cylinders in Chapter 4 were found. In general, the results based on the HOIBC solution have been found to be acceptable for dielectric coatings in the range less than one quarter wavelength deep. For magnetic coatings the allowable coating depth was found to be approximately half that for a dielectric coating. As expected, the accuracy of the HOIBC is poorest for bodies with sharp corners, and best when the radii of curvature on the BOR exceed one wavelength. The results for lossy coatings were in general better than those for lossless coatings, also as expected.

Properties of Exact Impedance Tensors

In this appendix several properties of the impedance tensor employed in Chapter 2 will be derived. In particular the cases of a lossless boundary and a reciprocal boundary are considered.

A.1 Impedance Matrices for Lossless Boundaries

Consider an impedance boundary located in the plane $z = 0$, as depicted in Figure A.1. An imaginary surface is placed at $z = 0^+$, just outside the boundary. The real power flow into the boundary is then given by [37]

$$P = \frac{1}{2}\text{Re}\left[\int_{-\infty}^{+\infty}\int_{-\infty}^{+\infty}(\mathbf{E} \times \mathbf{H}^*)\cdot(-\hat{\mathbf{z}})\,dxdy\right]. \qquad (A.1)$$

If the boundary is described by an impedance tensor as in Chapter 2, then

$$E_x = Z_{xx}H_x + Z_{xy}H_y \qquad (A.2)$$

and

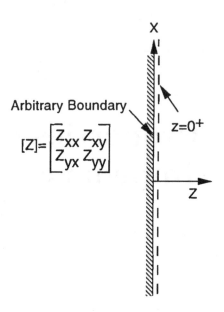

Figure A.1: A planar impedance boundary.

$$E_y = Z_{yx}H_x + Z_{yy}H_y. \qquad (A.3)$$

Substituting these expressions into Eq. (A.1),

$$P = \frac{1}{2}\text{Re}\left[\int_{-\infty}^{+\infty}\int_{-\infty}^{+\infty}\left(Z_{xx}H_xH_y^* + Z_{xy}\left|H_y\right|^2\right.\right.$$
$$\left.\left.-Z_{yx}\left|H_x\right|^2 - Z_{yy}H_x^*H_y\right)\cdot(-\mathbf{z})\,dxdy\right]. \qquad (A.4)$$

For a boundary that is neither lossy nor transparent $P = 0$, since there can be no net power flow into the surface. The above expression may be expanded to give

$$0 = \text{Re}\left[Z_{xy}\right]\left|H_y\right|^2 - \text{Re}\left[Z_{yx}\right]\left|H_x\right|^2$$
$$+\text{Re}\left[Z_{xx} - Z_{yy}\right]\text{Re}\left[H_xH_y^*\right] - \text{Im}\left[Z_{xx} + Z_{yy}\right]\text{Im}\left[H_xH_y^*\right]. \qquad (A.5)$$

Since the above must be true for all choices of H_x and H_y, we must have

$$\text{Re}\left[Z_{xy}\right] = \text{Re}\left[Z_{yx}\right] = 0, \qquad (A.6)$$

$$\text{Re}\left[Z_{xx}\right] = \text{Re}\left[Z_{yy}\right], \qquad (A.7)$$

and

$$\text{Im}\,[Z_{xx}] = -\text{Im}\,[Z_{yy}]\,, \tag{A.8}$$

or equivalently,

$$Z_{xx} = Z_{yy}^*. \tag{A.9}$$

These are the important relations between elements of the impedance tensor for an arbitrary lossless boundary. Examples include perfect conductors coated with layers of arbitrary lossless materials as well as treated conductors such as corrugated conductors.

A.2   Impedance Matrices for Reciprocal Boundaries

The relationships between the impedance terms when the boundary is reciprocal may be obtained by applying the reciprocity theorem [37] to the situation depicted in Figure A.1. If the region for $z < 0$ contains only reciprocal material then

$$\int_{-\infty}^{+\infty} \int_{-\infty}^{+\infty} \left(\mathbf{E}^{(1)} \times \mathbf{H}^{(2)} - \mathbf{E}^{(2)} \times \mathbf{H}^{(1)} \right) \cdot (-\mathbf{z})\, dx dy = 0 \tag{A.10}$$

for two arbitrary fields, $(\mathbf{E}^{(1)}, \mathbf{H}^{(1)})$ and $(\mathbf{E}^{(2)}, \mathbf{H}^{(2)})$ [37]. Substituting the relationships of Eqs. (A.2) and (A.3) into the above equation and canceling terms,

$$\int_{-\infty}^{+\infty} \int_{-\infty}^{+\infty} (Z_{xx} + Z_{yy}) \left(H_x^{(1)} H_y^{(2)} - H_x^{(2)} H_y^{(1)} \right) dx dy = 0. \tag{A.11}$$

Since this must be true for arbitrary choices of the fields, we obtain

$$Z_{xx} = -Z_{yy} \tag{A.12}$$

for a reciprocal boundary. It should be noted that Eqs. (A.12), (A.6), and (A.9) imply that all of the impedance terms are pure imaginary quantities for a lossless reciprocal boundary. In addition, the above proofs may be reproduced for the cylindrical boundaries of interest in Chapter 4. The results are identical to those for planar boundaries with the following substitutions: $Z_{xx} \rightarrow Z_{\phi\phi}$, $Z_{xy} \rightarrow Z_{\phi z}$, $Z_{yx} \rightarrow Z_{z\phi}$, and $Z_{yy} \rightarrow Z_{zz}$.

A.2 Impedance Matrices for Reciprocal Boundaries

Appendix B

Symmetry Properties of the Polynomials $P_1 - P_8$

In this appendix symmetry relations for the polynomials discussed in Chapter 2 will be derived. Consider the planar boundary depicted in Figure B.1. In the original (x, y, z) coordinate system the approximate spectral domain boundary

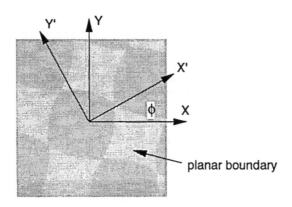

Figure B.1: A planar boundary with symmetry under a rotation angle of ϕ.

139

condition is given by,

$$\begin{bmatrix} P_1\left(k_x, k_y\right) & P_2\left(k_x, k_y\right) \\ P_3\left(k_x, k_y\right) & P_4\left(k_x, k_y\right) \end{bmatrix} \begin{bmatrix} \tilde{E}_x\left(k_x, k_y\right) \\ \tilde{E}_y\left(k_x, k_y\right) \end{bmatrix}$$

$$= \begin{bmatrix} P_5\left(k_x, k_y\right) & P_6\left(k_x, k_y\right) \\ P_7\left(k_x, k_y\right) & P_8\left(k_x, k_y\right) \end{bmatrix} \begin{bmatrix} \tilde{H}_x\left(k_x, k_y\right) \\ \tilde{H}_y\left(k_x, k_y\right) \end{bmatrix}. \tag{B.1}$$

Now suppose this boundary condition is expressed in the rotated coordinate system, (x', y', z'), with $z = z'$,

$$\begin{bmatrix} P_1\left(k_x', k_y'\right) & P_2\left(k_x', k_y'\right) \\ P_3\left(k_x', k_y'\right) & P_4\left(k_x', k_y'\right) \end{bmatrix} \begin{bmatrix} \cos\phi & -\sin\phi \\ \sin\phi & \cos\phi \end{bmatrix} \begin{bmatrix} \tilde{E}_{x'}\left(k_x', k_y'\right) \\ \tilde{E}_{y'}\left(k_x', k_y'\right) \end{bmatrix}$$

$$= \begin{bmatrix} P_5\left(k_x', k_y'\right) & P_6\left(k_x', k_y'\right) \\ P_7\left(k_x', k_y'\right) & P_8\left(k_x', k_y'\right) \end{bmatrix} \begin{bmatrix} \cos\phi & -\sin\phi \\ \sin\phi & \cos\phi \end{bmatrix} \begin{bmatrix} \tilde{H}_{x'}\left(k_x', k_y'\right) \\ \tilde{H}_{y'}\left(k_x', k_y'\right) \end{bmatrix} \tag{B.2}$$

The polynomials are now expressed in terms of the primed wavenumbers, which are related to the unprimed quantities through

$$\begin{aligned} k_x &= k_x' \cos\phi - k_y' \sin\phi \\ k_y &= k_x' \sin\phi + k_y' \cos\phi. \end{aligned} \tag{B.3}$$

If the boundary is invariant under the rotation ϕ then the transformed boundary condition, Eq. (B.2), must be indistinguishable from the original one, Eq. (B.1). Each of the two new equations may, however, appear as a linear combination of the two original equations. This condition is expressible in matrix form as,

$$\begin{bmatrix} P_1\left(k_x', k_y'\right) & P_2\left(k_x', k_y'\right) \\ P_3\left(k_x', k_y'\right) & P_4\left(k_x', k_y'\right) \end{bmatrix} \begin{bmatrix} \cos\phi & -\sin\phi \\ \sin\phi & \cos\phi \end{bmatrix}$$

$$= \begin{bmatrix} \alpha & \beta \\ \gamma & \delta \end{bmatrix} \begin{bmatrix} P_1\left(k_x, k_y\right) & P_2\left(k_x, k_y\right) \\ P_3\left(k_x, k_y\right) & P_4\left(k_x, k_y\right) \end{bmatrix}. \tag{B.4}$$

The unknown constants α, β, γ, and δ are used to form the linear combination of the original boundary conditions and will be determined next. An identical equation holds for the right-hand matrix of polynomials, and it may be obtained from the above expression when $P_1 \to P_5$, $P_2 \to P_6$, $P_3 \to P_7$, and $P_4 \to P_8$.

In the development of the original boundary condition it is always possible to choose the constant coefficients in the polynomials $P_2\left(k_x, k_y\right)$ and $P_3\left(k_x, k_y\right)$ to be zero, and the constant coefficients of the polynomials $P_1\left(k_x, k_y\right)$ and $P_4\left(k_x, k_y\right)$ to be equal to one. Under these circumstances we may evaluate Eq. (B.4) for the case $k_x = k_y = 0$ and obtain the unknown coefficients immediately as

$$\begin{aligned} \alpha &= \cos\phi, \\ \beta &= -\sin\phi, \\ \gamma &= \sin\phi, \end{aligned} \tag{B.5}$$

and

$$\delta = \cos\phi. \tag{B.6}$$

The conditions on the polynomials for a boundary that is invariant under a rotation ϕ are then given by

$$\begin{bmatrix} P_1\left(k_x',k_y'\right) & P_2\left(k_x',k_y'\right) \\ P_3\left(k_x',k_y'\right) & P_4\left(k_x',k_y'\right) \end{bmatrix} \begin{bmatrix} \cos\phi & -\sin\phi \\ \sin\phi & \cos\phi \end{bmatrix}$$

$$= \begin{bmatrix} \cos\phi & -\sin\phi \\ \sin\phi & \cos\phi \end{bmatrix} \begin{bmatrix} P_1\left(k_x,k_y\right) & P_2\left(k_x,k_y\right) \\ P_3\left(k_x,k_y\right) & P_4\left(k_x,k_y\right) \end{bmatrix} \tag{B.7}$$

and

$$\begin{bmatrix} P_5\left(k_x',k_y'\right) & P_6\left(k_x',k_y'\right) \\ P_7\left(k_x',k_y'\right) & P_8\left(k_x',k_y'\right) \end{bmatrix} \begin{bmatrix} \cos\phi & -\sin\phi \\ \sin\phi & \cos\phi \end{bmatrix}$$

$$= \begin{bmatrix} \cos\phi & -\sin\phi \\ \sin\phi & \cos\phi \end{bmatrix} \begin{bmatrix} P_5\left(k_x,k_y\right) & P_6\left(k_x,k_y\right) \\ P_7\left(k_x,k_y\right) & P_8\left(k_x,k_y\right) \end{bmatrix}. \tag{B.8}$$

When the boundary is invariant under a 180 degree rotation the above equations require that the linear terms k_x and k_y in all of the polynomials must vanish. For the case of a boundary with 90 degree symmetry we obtain

$$P_1(-k_y, k_x) = P_4(k_x, k_y)$$
$$P_5(-k_y, k_x) = P_8(k_x, k_y)$$
$$P_3(-k_y, k_x) = -P_2(k_x, k_y)$$
$$P_6(-k_y, k_x) = -P_7(k_x, k_y). \tag{B.9}$$

These relations can be used to reduce the total number of coefficients to be determined from 48 to 14 for a boundary with 180 degree and 90 degree symmetry. If the condition of arbitrary ϕ is enforced, i.e., there is complete rotational symmetry, four more coefficients may be determined, resulting in a total of 10 unknown coefficients and the form of the polynomials given in Eqs. (2.19) to (2.22).

Appendix C

Surface Waves on Impedance Surfaces

In this appendix the important topic of surface waves on surfaces whose behavior may be described using an impedance tensor will be addressed. Consider the general planar boundary depicted earlier in Figure A.1. We assume that the surface and therefore the impedance tensor are invariant under an arbitrary rotation. In this case it is sufficient to consider the case of propagation in the x direction only, i.e., $k_y = 0$. We wish to determine the possible surface wave modes of the planar boundary.

Consider the region $z = 0^+$. The field in this exterior region may be expressed as a sum of TE and TM to z fields [37]. The vector potential describing these fields is a decaying function in the z direction and a propagating wave in the x direction. If the vector potential is denoted as Ψ, then we have,

$$\Psi = e^{-jk_x x - \alpha z}, \tag{C.1}$$

where the attenuation coefficient α and propagation coefficient k_x are related to the free space wavenumber k_0 through

$$k_0^2 = k_x^2 - \alpha^2. \tag{C.2}$$

Substituting the expression for Ψ into the equations in Ref. [37] we find the following nonzero field components for the TE to z mode,

$$E_y(k_x) = -jk_x\Psi \tag{C.3}$$

$$H_x(k_x) = \frac{k_x\alpha}{\omega\mu_0}\Psi \tag{C.4}$$

$$H_z(k_x) = \frac{k_x^2}{j\omega\mu_0}\Psi. \tag{C.5}$$

Similarly, using the same expression for Ψ the nonzero components for the TM to z field are found,

$$E_x(k_x) = \frac{k_x\alpha}{\omega\epsilon_0}\Psi \tag{C.6}$$

$$H_y(k_x) = jk_x\Psi \tag{C.7}$$

$$E_z(k_x) = \frac{k_x^2}{j\omega\epsilon_0}\Psi. \tag{C.8}$$

Using the above expressions we may define an impedance tensor for the surface wave mode, determined only from information in the region $z > 0$,

$$\left\{ \begin{array}{c} E_x(k_x) \\ E_y(k_x) \end{array} \right\} = \left[\begin{array}{cc} Z_{xx}^{sw}(k_x) & Z_{xy}^{sw}(k_x) \\ Z_{yx}^{sw}(k_x) & Z_{yy}^{sw}(k_x) \end{array} \right] \left\{ \begin{array}{c} H_x(k_x) \\ H_y(k_x) \end{array} \right\}, \tag{C.9}$$

or

$$\{E\} = [Z^{sw}]\{H\}. \tag{C.10}$$

The individual terms of the impedance tensor are then given by,

$$Z_{xx}^{sw}(k_x) = 0 \tag{C.11}$$

$$Z_{xy}^{sw}(k_x) = \frac{-j\sqrt{k_x^2 - k_0^2}}{\omega\epsilon_0} \tag{C.12}$$

$$Z_{yx}^{sw}(k_x) = \frac{-j\omega\mu_0}{\sqrt{k_x^2 - k_0^2}} \tag{C.13}$$

$$Z_{yy}^{sw}(k_x) = 0. \tag{C.14}$$

The impedance properties of the boundary also require that

$$\{E\} = [Z]\{H\}. \tag{C.15}$$

When Eqs. (C.10) and (C.15) are combined we obtain

$$\{0\} = ([Z] - [Z^{sw}])\{H\}. \tag{C.16}$$

Surface waves exist for values of k_x that allow a nontrivial solution of this equation. Thus the surface wave values for k_x are given by

$$\text{Det}\left[[Z] - [Z^{sw}]\right] = 0. \tag{C.17}$$

Substituting in the expressions for the terms of $[Z^{sw}]$ the final characteristic equation is obtained,

$$Z_{xx}(k_x)Z_{yy}(k_x) - \left(Z_{yx}(k_x) + \frac{j\omega\mu_0}{\sqrt{k_x^2 - k_0^2}}\right)\left(Z_{xy}(k_x) + \frac{j\sqrt{k_x^2 - k_0^2}}{\omega\epsilon_0}\right) = 0.$$

(C.18)

As an application of the general characteristic equation, Eq. (C.18), consider the case of a dielectric layer. In this case the appropriate impedance tensor is given in Chapter 3, Eqs. (3.5) and (3.6). In this particular case the impedances Z_{xx} and Z_{yy} are zero, and the characteristic equation decouples into two independent equations,

$$\frac{-\sqrt{k^2 - k_x^2}}{\sqrt{k_x^2 - k_0^2\mu_r}} = \tan\sqrt{k^2 - k_x^2}d$$

(C.19)

and

$$\frac{-\sqrt{k_x^2 - k_0^2\epsilon_r}}{\sqrt{k^2 - k_x^2}} = \tan\sqrt{k^2 - k_x^2}d,$$

(C.20)

where d is the depth of the layer, $\epsilon_r = \epsilon/\epsilon_0$, and $\mu_r = \mu/\mu_0$. Equations (C.19) and (C.20) may be recognized as the well known characteristic equations for the TE and TM surface wave modes for a dielectric layer, which have been derived previously using many different approaches.

The present development is quite general and is applicable for planar boundaries with arbitrary $[Z]$. When Z_{xx} and Z_{yy} are nonzero when $k_y = 0$, as is the case for a chiral layer (Chapter 3), then the TE and TM surface wave modes become coupled. Such behavior has already been described for the chiral layer [47]. In this case the wavenumbers for the surface wave modes lie between k_0 and the maximum of the right-handed and left-handed wavenumbers, k_r and k_l, for the chiral medium.

The above development may also be used to determine the ability of the HOIBC to correctly predict surface wave behavior. When HOIBC are used to model a boundary then the approximate impedance tensor, i.e., the ratio of polynomial approximation, must be substituted into Eq. (C.18) rather than the exact impedance tensor. If the HOIBC is accurate for k_x in the vicinity of the surface wave then the roots of the characteristic equation will be essentially the same and the surface waves will also be predicted with high accuracy. On the other hand if the HOIBC accuracy tails off significantly in the surface wave region then surface wave behavior will not be predicted well. As was pointed out in Chapter 3, this is the reason for checking the accuracy of the HOIBC not only for values of k_x in the visible range but in the surface wave region as well. This ability to examine the impedance boundary condition in the surface wave region in a straightforward manner is one of the key advantages of the spectral domain approach.

Appendix D

Plane Wave Scattering at an Impedance Plane

In this appendix solutions for plane wave scattering at a planar interface, which is modeled by either the exact spectral domain impedance boundary condition or a higher order impedance boundary condition, will be derived. Results based upon these equations are then used to assess the accuracy of the higher order impedance boundary conditions in Chapter 3.

Consider the problem of plane wave reflection at a planar boundary, depicted in Figure D.1. Again we assume the coating is rotationally invariant and consider the case $k_y = 0$. An external incident field that is either TE to z (H polarization) with amplitude A_{TE} or TM to z (E polarization) with amplitude A_{TM} strikes the plane, generating two reflected waves, one TE to z with amplitude B_{TE}, and one TM to z with amplitude B_{TM}. Given matrices M_E and M_H, which relate tangential field components at $z = 0$ to the wave amplitudes, and the impedance tensor for the coated plane $[Z]$, the reflection coefficients may be calculated for each wavenumber k_x as follows. If, at $z = 0$,

$$\{E\} = \left\{ \begin{array}{c} E_x \\ E_y \end{array} \right\}, \ \{H\} = \left\{ \begin{array}{c} H_x \\ H_y \end{array} \right\} \tag{D.1}$$

and

$$\{A\} = \left\{ \begin{array}{c} A_{TE} \\ A_{TM} \end{array} \right\} e^{-jk_x x}, \ \{B\} = \left\{ \begin{array}{c} B_{TE} \\ B_{TM} \end{array} \right\} e^{-jk_x x}, \tag{D.2}$$

147

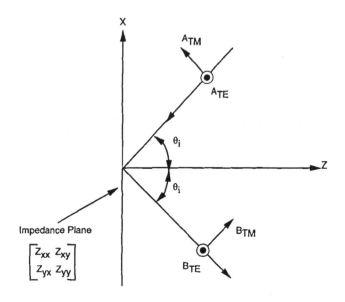

Figure D.1: Plane wave reflection at a planar boundary.

then

$$\{E\} = [M_E]\left(\{A\} + \{B\}\right) \tag{D.3}$$

and

$$\{H\} = [M_H]\left(\{A\} - \{B\}\right), \tag{D.4}$$

with

$$[M_E] = \begin{bmatrix} 0 & k_z/k \\ 1 & 0 \end{bmatrix}, \ [M_H] = \frac{1}{\eta_0}\begin{bmatrix} k_z/k & 0 \\ 0 & -1 \end{bmatrix}, \tag{D.5}$$

and $k_z = \sqrt{k_0^2 - k_x^2}$, where k_0 is the free space wavenumber. For a given k_x and corresponding impedance tensor

$$[Z] = \begin{bmatrix} Z_{xx} & Z_{xy} \\ Z_{yx} & Z_{yy} \end{bmatrix}, \tag{D.6}$$

the reflected field components are found using $\{E\} = [Z]\{H\}$. The reflection coefficient $[R]$ is defined according to $\{B\} = [R]\{A\}$, with individual components given by

$$[R] = \begin{bmatrix} R_{TETE} & R_{TETM} \\ R_{TMTE} & R_{TMTM} \end{bmatrix}. \tag{D.7}$$

Using the previous relationships the reflection coefficient matrix is found to be

$$[R] = -\left([M_E] + [Z][M_H]\right)^{-1}\left([M_E] - [Z][M_H]\right). \tag{D.8}$$

Since the matrices $[M_E]$, $[M_H]$, and $[Z]$ are in general all functions of k_x, so are the reflection coefficients of the boundary.

For a rigorous solution the impedance terms in the above equation are evaluated exactly for each value of k_x and the reflection coefficients are computed. Exact impedance matrices for a dielectric coating, chiral coating, and corrugated conductor are given in Chapter 3. When the Tensor Impedance Boundary Condition, TIBC, or the Standard Impedance Boundary Condition, SIBC, is used to approximate the boundary's behavior then the impedance tensor $[Z]$ is assumed to be equal to its value for normal incidence, $k_x = 0$, for all values of k_x. When a Higher Order Impedance Boundary Condition, HOIBC, is used to model the boundary's behavior the impedance tensor is given by the ratio of polynomial approximation of Eq. (2.18).

Appendix E

Plane Wave Scattering
at an Impedance Cylinder

In this appendix solutions for plane wave scattering at a circular cylinder, which is modeled by either the exact spectral domain impedance boundary condition or a higher order impedance boundary condition, will be derived. Results based upon these equations are then used to assess the accuracy of the higher order impedance boundary conditions in Chapter 4.

The solution for plane wave scattering by the arbitrary circular cylinder of radius b depicted in Figure E.1 proceeds by expanding the incident plane wave into a Fourier series in ϕ. For a TM to z (E polarization) plane wave incident in the x, $\phi = 0$ direction [37],

$$E_z^i = \sum_{n=-\infty}^{+\infty} j^{-n} J_n(k_0 b) e^{jn\phi}, \tag{E.1}$$

where J_n is the Bessel function of the first kind of order n. Likewise for a TE to z (H polarization) plane wave incident in the x, $\phi = 0$ direction,

$$H_z^i = \frac{j}{\eta_0} \sum_{n=-\infty}^{+\infty} j^{-n} J_n(k_0 b) e^{jn\phi}, \tag{E.2}$$

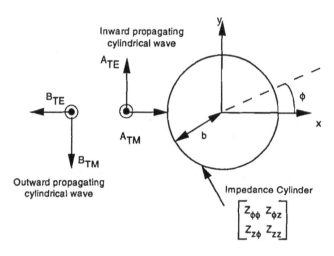

Figure E.1: Plane wave scattering by a circular cylinder.

where η_0 is the free space wave impedance. Due to the orthogonality of the terms in the Fourier series the reflected field for each Fourier component may be obtained independently in a manner analogous to that employed in the planar case, Appendix D. If, at $r = b$

$$\{E\} = \left\{ \begin{array}{c} E_\phi \\ E_z \end{array} \right\}, \ \{H\} = \left\{ \begin{array}{c} H_\phi \\ H_z \end{array} \right\} \tag{E.3}$$

and

$$\{A\} = \left\{ \begin{array}{c} A_{TE} \\ A_{TM} \end{array} \right\}, \ \{B\} = \left\{ \begin{array}{c} B_{TE} \\ B_{TM} \end{array} \right\}, \tag{E.4}$$

then the vectors $\{E\}$ and $\{H\}$ may be written in terms of the TE to z and TM to z incident and reflected cylindrical waves according to

$$\{E\} = \left[M_E^A \right] \{A\} + \left[M_E^B \right] \{B\} \tag{E.5}$$

and

$$\{H\} = \left[M_H^A \right] \{A\} + \left[M_H^B \right] \{B\} \tag{E.6}$$

where

$$\left[M_E^A \right] = j^{-n} \left[\begin{array}{cc} -J_n'(k_0 b) & 0 \\ 0 & J_n(k_0 b) \end{array} \right], \tag{E.7}$$

$$\left[M_E^B \right] = j^{-n} \left[\begin{array}{cc} -H_n^{(2)\prime}(k_0 b) & 0 \\ 0 & H_n^{(2)}(k_0 b) \end{array} \right], \tag{E.8}$$

$$\left[M_H^A \right] = \frac{j^{-n}}{\eta_0} \left[\begin{array}{cc} 0 & -j J_n'(k_0 b) \\ j J_n(k_0 b) & 0 \end{array} \right], \tag{E.9}$$

and

$$[M_H^B] = \frac{j^{-n}}{\eta_0} \begin{bmatrix} 0 & -jH_n^{(2)\prime}(k_0b) \\ jH_n^{(2)}(k_0b) & 0 \end{bmatrix}. \tag{E.10}$$

Once again the scattered field components are found using $\{E\} = [Z]\{H\}$. The reflection coefficient matrix $[R]$ for the nth term is defined according to $\{B\} = [R]\{A\}$, with individual components given by,

$$[R] = \begin{bmatrix} R_{TETE} & R_{TETM} \\ R_{TMTE} & R_{TMTM} \end{bmatrix}. \tag{E.11}$$

The reflection coefficient matrix is then given for the nth term in the Fourier series by

$$[R] = -\left([M_E^B] - [Z][M_H^B]\right)^{-1}\left([M_E^A] - [Z][M_H^A]\right). \tag{E.12}$$

This equation is used to compute the reflected TE to z and TM to z components for an arbitrary incident field. If the incident field is TE to z then

$$\{A\} = \left\{ \begin{array}{c} 1 \\ 0 \end{array} \right\}, \tag{E.13}$$

and if the incident field is TM to z then

$$\{A\} = \left\{ \begin{array}{c} 0 \\ 1 \end{array} \right\}, \tag{E.14}$$

for each value of n.

In general the impedance tensor $[Z]$ is a function of n. Formulas for computing this tensor for dielectric-coated and chiral-coated conducting cylinders are given in Chapter 4. When the TIBC or SIBC is employed the tensor $[Z]$ is assumed to be independent of n and equal to its value when $n = 0$. For the HOIBC solution the impedance tensor is replaced by its ratio of polynomial approximation, Eq. (2.18), with the polynomials given by Eqs. (4.19) to (4.22), evaluated for $k_t = n/b$.

Matrix Elements for the CFIE Portion of the BOR Solution

In this appendix detailed expressions for the matrix elements in the Combined Field Integral Equation (CFIE) for the exterior region will be presented.

The two matrix equations obtained by Galerkin's method are given by

$$V_n^t(i) = \sum_{j=1}^{J_T} Z_J^{tt}(i,j)a_n^t(j)$$

$$+ Z_J^{t\phi}(i,j)a_n^\phi(j) + Z_M^{tt}(i,j)b_n^t(j) + Z_M^{t\phi}(i,j)b_n^\phi(j) \tag{F.1}$$

and

$$V_\phi(i) = \sum_{j=1}^{J_T} Z_J^{\phi t}(i,j)a_n^t(j)$$

$$+ Z_J^{\phi\phi}(i,j)a_n^\phi(j) + Z_M^{\phi t}(i,j)b_n^t(j) + Z_M^{\phi\phi}(i,j)b_n^\phi(j). \tag{F.2}$$

The impedance matrix elements may be written in terms of the operators K and L and a third operator $\tilde{K}$, which are defined in Ref. [7].

155

$$Z_J^{tt}(i,j) = L^{tt}(i,j) + \frac{\alpha}{1-\alpha}\tilde{K}^{tt}(i,j) + \frac{\alpha}{1-\alpha}K^{\phi t}(i,j), \qquad (F.3)$$

$$Z_J^{t\phi}(i,j) = L^{t\phi}(i,j) + \frac{\alpha}{1-\alpha}K^{\phi\phi}(i,j), \qquad (F.4)$$

$$Z_J^{\phi t}(i,j) = L^{\phi t}(i,j) - \frac{\alpha}{1-\alpha}K^{t\phi}(i,j), \qquad (F.5)$$

$$Z_J^{\phi\phi}(i,j) = L^{\phi\phi}(i,j) + \frac{\alpha}{1-\alpha}\tilde{K}^{\phi\phi}(i,j) + \frac{\alpha}{1-\alpha}K^{t\phi}(i,j), \qquad (F.6)$$

$$Z_M^{tt}(i,j) = \frac{1}{\eta_0}\left(-K^{tt}(i,j) + \frac{\alpha}{1-\alpha}L^{\phi t}(i,j)\right), \qquad (F.7)$$

$$Z_M^{t\phi}(i,j) = \frac{1}{\eta_0}\left(-K^{t\phi}(i,j) + \tilde{K}^{\phi\phi}(i,j) + \frac{\alpha}{1-\alpha}L^{\phi\phi}(i,j)\right), \qquad (F.8)$$

$$Z_M^{\phi t}(i,j) = \frac{1}{\eta_0}\left(-K^{\phi t}(i,j) - \tilde{K}^{tt}(i,j) - \frac{\alpha}{1-\alpha}L^{tt}(i,j)\right), \qquad (F.9)$$

and

$$Z_M^{\phi\phi}(i,j) = \frac{1}{\eta_0}\left(-K^{\phi\phi}(i,j) - \frac{\alpha}{1-\alpha}L^{t\phi}(i,j)\right). \qquad (F.10)$$

In obtaining the elements of the submatrices the integrals in the t direction along each segment of the triangle are approximated as being equal to the value of the integrand evaluated at the center of the segment times the length of the segment. The ϕ integrations are performed numerically using Gaussian quadrature. The elements of the K matrices are given by

$$K^{tt}(i,j) = -j\eta_0 \sum_{p=2i-1}^{2i+2}\sum_{q=2j-1}^{2j+2} ((z_p - z_q)\sin\gamma_p\sin\gamma_q$$
$$+\rho_q\sin\gamma_p\cos\gamma_q - \rho_p\cos\gamma_p\sin\gamma_q) T_p^i T_q^j I_{pq}^{ss}, \qquad (F.11)$$

$$K^{\phi t}(i,j) = \eta_0 \sum_{p=2i-1}^{2i+2}\sum_{q=2j-1}^{2j+2} (((z_p - z_q)\sin\gamma_q$$
$$+\rho_q\cos\gamma_q I_{pq}^{cc} - \rho_p\cos\gamma_q I_{pq}^c) T_p^i T_q^j, \qquad (F.12)$$

$$K^{t\phi}(i,j) = -\eta_0 \sum_{p=2i-1}^{2i+2}\sum_{q=2j-1}^{2j+2} ((-(z_p - z_q)\sin\gamma_p$$
$$+\rho_p\cos\gamma_p I_{pq}^{cc} - \rho_q\cos\gamma_p I_{pq}^c) T_p^i T_q^j, \qquad (F.13)$$

and

$$K^{\phi\phi}(i,j) = j\eta_0 \sum_{p=2i-1}^{2i+2} \sum_{q=2j-1}^{2j+2} (z_p - z_q) T_p^i T_q^j I_{pq}^{ss}, \qquad (F.14)$$

where

$$I_{pq}^c = l_p l_q \int_0^\pi \frac{(1 + jk_0 R_{pq})}{R_{pq}^3} \cos n\phi\, e^{-jk_0 R_{pq}} d\phi, \qquad (F.15)$$

$$I_{pq}^{cc} = l_p l_q \int_0^\pi \frac{(1 + jk_0 R_{pq})}{R_{pq}^3} \cos n\phi \cos\phi\, e^{-jk_0 R_{pq}} d\phi, \qquad (F.16)$$

and

$$I_{pq}^{ss} = l_p l_q \int_0^\pi \frac{(1 + jk_0 R_{pq})}{R_{pq}^3} \sin n\phi \sin\phi\, e^{-jk_0 R_{pq}} d\phi. \qquad (F.17)$$

The elements of the L matrices are given by

$$L^{tt}(i,j) = \sum_{p=2i-1}^{2i+2} \sum_{q=2j-1}^{2j+2} j\omega\mu_0 T_p^i T_q^j \left(\sin\gamma_p \sin\gamma_q K_{pq}^{cc} \right.$$
$$\left. + \cos\gamma_p \cos\gamma_q K_{pq}^c \right) - \frac{j}{\omega\epsilon_0} \dot{T}_p^i \dot{T}_q^j K_{pq}^c, \qquad (F.18)$$

$$L^{\phi t}(i,j) = - \sum_{p=2i-1}^{2i+2} \sum_{q=2j-1}^{2j+2} \omega\mu_0 \sin\gamma_q T_p^i T_q^j K_{pq}^{ss} + \frac{n}{\omega\epsilon_0 \rho_p} T_p^i \dot{T}_q^j K_{pq}^c, \qquad (F.19)$$

$$L^{t\phi}(i,j) = - \sum_{p=2i-1}^{2i+2} \sum_{q=2j-1}^{2j+2} \omega\mu_0 \sin\gamma_p T_p^i T_q^j K_{pq}^{ss} + \frac{n}{\omega\epsilon_0 \rho_q} \dot{T}_p^i T_q^j K_{pq}^c, \qquad (F.20)$$

and

$$L^{\phi\phi}(i,j) = - \sum_{p=2i-1}^{2i+2} \sum_{q=2j-1}^{2j+2} \left(j\omega\mu_0 K_{pq}^{cc} + \frac{n^2}{j\omega\epsilon_0 \rho_p \rho_q} K_{pq}^c \right) T_p^i T_q^j, \qquad (F.21)$$

where

$$K_{pq}^c = l_p l_q \int_0^\pi \frac{e^{-jk_0 R_{pq}}}{R_{pq}} \cos n\phi\, d\phi, \qquad (F.22)$$

$$K_{pq}^{cc} = l_p l_q \int_0^\pi \frac{e^{-jk_0 R_{pq}}}{R_{pq}} \cos n\phi \cos\phi\, d\phi, \qquad (F.23)$$

$$K_{pq}^{ss} = l_p l_q \int_0^\pi \frac{e^{-jk_0 R_{pq}}}{R_{pq}} \sin n\phi \sin \phi \, d\phi, \tag{F.24}$$

and

$$R_{pq} = \sqrt{(\rho_p - \rho_q)^2 + (z_p - z_q)^2 + 2\rho_p\rho_q(1 - \cos\phi)}, \quad p \neq q, \tag{F.25}$$

or

$$R_{pq} = \sqrt{\left(\frac{l_p}{4}\right)^2 + 2\rho_p^2(1 - \cos\phi)}, \quad p = q. \tag{F.26}$$

The function ρ_p is the radius to the center of the pth segment and T_p^i is the value of the ith triangle function at the center of the pth segment. The notation $\dot{T}_p^i$ denotes the derivative of the triangle function, and l_p is the length of the pth segment. Here γ_p is the angle of the pth segment relative to the z axis.

The matrices $\tilde{K}^{tt}$ and $\tilde{K}^{\phi\phi}$ represent the self terms. Their entries are given by

$$\tilde{K}^{tt}(i,j) = -\tilde{K}^{\phi\phi}(i,j) = \pi\eta_0 \sum_{p=2i-1}^{2i+2} \sum_{q=2j-1}^{2j+2} l_p \frac{T_p^i T_q^j}{\rho_p} \delta_{pq}, \tag{F.27}$$

where δ_{pq} indicates the Kroneker delta function. The entries of the source vectors V^t and V^ϕ are obtained by integrating the dot product of the testing functions with the incident fields over the exterior surface of the BOR,

$$V_t(i) = -\int_{S_0} \mathbf{W}^t(i) \cdot \left[\hat{a}_n^0 \times \mathbf{E}^i(\mathbf{r}) + \frac{\alpha}{1-\alpha} \left(\hat{a}_n^0 \times \hat{a}_n^0 \times \mathbf{H}^i(\mathbf{r}) \right) \right] ds \tag{F.28}$$

and

$$V_\phi(i) = -\int_{S_0} \mathbf{W}^\phi(i) \cdot \left[\hat{a}_n^0 \times \mathbf{E}^i(\mathbf{r}) + \frac{\alpha}{\alpha-1} \left(\hat{a}_n^0 \times \hat{a}_n^0 \times \mathbf{H}^i(\mathbf{r}) \right) \right] ds. \tag{F.29}$$

The surface integrations required to determine the source vectors can be carried out in a manner similar to that used for the terms of the impedance matrix. In the case of an incident plane wave the ϕ integrations may be computed analytically.

Transformation of the HOIBC onto the BOR

In this appendix the unit vectors and transformation rules for the differential operators appearing in Eqs. (7.24) and (7.25) are derived. The pertinent geometry is shown in Figure 7.5. In terms of the t, ϕ components the position of any point on the BOR is given by,

$$\mathbf{r}(t,\phi) = (x(t,\phi), y(t,\phi), z(t,\phi)) = (\rho(t)\cos\phi, \rho(t)\sin\phi, z(t)). \tag{G.1}$$

The unit vectors $\hat{\mathbf{t}}$ and $\hat{\boldsymbol{\phi}}$ are given by

$$\hat{\mathbf{t}}(t,\phi) = \left(\frac{\partial\rho}{\partial t}\cos\phi, \frac{\partial\rho}{\partial t}\sin\phi, \frac{\partial z}{\partial t}\right) \tag{G.2}$$

and

$$\hat{\boldsymbol{\phi}}(t,\phi) = (-\sin\phi, \cos\phi, 0). \tag{G.3}$$

Unit vectors in the $\hat{\mathbf{a}}_1$ and $\hat{\mathbf{a}}_2$ directions may be found from

$$\hat{\mathbf{a}}_1 \cdot \hat{\mathbf{y}}' = \hat{\mathbf{a}}_1 \cdot \hat{\mathbf{t}}\big|_{P_0} = 0 \tag{G.4}$$

and

$$\hat{a}_2 \cdot \hat{x}' = \hat{a}_1 \cdot \hat{\phi}\big|_{P_0} = 0, \tag{G.5}$$

where P_0 indicates the unit vectors are to be evaluated at the point $l_1 = l_2 = 0$, i.e., $t = t_0$ and $\phi = 0$. These vectors are given by

$$\hat{t}\big|_{P_0} = \left(\frac{\partial \rho}{\partial t}\bigg|_{P_0}, 0, \frac{\partial z}{\partial t}\bigg|_{P_0} \right) \tag{G.6}$$

and

$$\hat{\phi}\big|_{P_0} = (0, 1, 0). \tag{G.7}$$

Using the above expressions the unit vectors are found to be

$$\hat{a}_1(t, \phi) = \left(\left(\frac{\partial \rho}{\partial t}\bigg|_{P_0} \frac{\partial \rho}{\partial t} \cos \phi + \frac{\partial z}{\partial t}\bigg|_{P_0} \frac{\partial z}{\partial t} \right) \hat{\phi} + \frac{\partial \rho}{\partial t}\bigg|_{P_0} \sin \phi \hat{t} \right) / D_1(t, \phi), \tag{G.8}$$

with

$$D_1(t, \phi) = \sqrt{ \left(\frac{\partial \rho}{\partial t}\bigg|_{P_0} \frac{\partial \rho}{\partial t} \cos \phi + \frac{\partial z}{\partial t}\bigg|_{P_0} \frac{\partial z}{\partial t} \right)^2 + \left(\frac{\partial \rho}{\partial t}\bigg|_{P_0} \sin \phi \right)^2 }, \tag{G.9}$$

and

$$\hat{a}_2(t, \phi) = \cos \phi \hat{t} - \frac{\partial \rho}{\partial t} \sin \phi \hat{\phi} / D_2(t, \phi), \tag{G.10}$$

with

$$D_2(t, \phi) = \sqrt{ \cos^2 \phi + \left(\frac{\partial \rho}{\partial t} \right)^2 \sin^2 \phi }. \tag{G.11}$$

These unit vectors may be used to find the vector components V_1, V_2 in terms of the components V_t, V_ϕ at any point on the BOR.

The next step is to transform the differential operators into the t, ϕ coordinate system. We begin by transferring the $\partial/\partial l_2$ operator which is particularly simple. In general the operator may be written as

$$\frac{\partial}{\partial l_2} = \left(\frac{dt}{dl_2} \right) \left(\frac{\partial}{\partial t} + \frac{d\phi}{dt} \frac{\partial}{\partial \phi} \right), \tag{G.12}$$

where it has been assumed that in general t and ϕ are not independent on the l_2 curve. In this particular case ϕ is a constant on any l_2 curve, and hence independent of t, and $dt/dl_2 = 1$, and the correct transformation is

$$\frac{\partial}{\partial l_2}\bigg|_{P_0} = \frac{\partial}{\partial t}. \tag{G.13}$$

The second derivative follows easily,

$$\frac{\partial^2}{\partial l_2^2}\bigg|_{P_0} = \frac{\partial^2}{\partial t^2}. \tag{G.14}$$

Now consider the $\partial/\partial l_1$ operator. In general the operator may be written as

$$\frac{\partial}{\partial l_1} = \left(\frac{d\phi}{dl_1}\right)\left(\frac{\partial}{\partial \phi} + \frac{dt}{d\phi}\frac{\partial}{\partial t}\right) \tag{G.15}$$

where again it has been assumed that in general t and ϕ are not independent on the l_1 curve. The expression for the second derivative is

$$\frac{\partial^2}{\partial l_1^2} = \left(\frac{d^2\phi}{dl_1^2}\right)\left(\frac{\partial}{\partial \phi} + \frac{dt}{d\phi}\frac{\partial}{\partial t}\right)$$
$$+ \left(\frac{d\phi}{dl_1}\right)^2\left(\frac{\partial^2}{\partial \phi^2} + \frac{d^2t}{d\phi^2}\frac{\partial}{\partial t} + 2\frac{dt}{d\phi}\frac{\partial^2}{\partial t\partial \phi} + \left(\frac{dt}{d\phi}\right)^2\frac{\partial^2}{\partial t^2}\right). \tag{G.16}$$

In this particular case ϕ and t are not independent on the l_1 curve. To find the relationship between t and ϕ on the l_1 curve passing through the point P_0, we note that points on this curve must satisfy

$$(\mathbf{r} - \mathbf{r}_0)\cdot\hat{\mathbf{t}}_0 = 0, \tag{G.17}$$

where the vector $\mathbf{r}_0$ points to the point P_0. Substituting in expressions for the vectors we obtain an implicit relationship between t and ϕ,

$$F(t,\phi) = \frac{\partial \rho}{\partial t}\bigg|_{P_0}(\rho(t)\cos\phi - \rho_0) + \frac{\partial z}{\partial t}\bigg|_{P_0}(z(y) - z_0) = 0. \tag{G.18}$$

To find the required derivatives we use implicit differentiation,

$$\frac{\partial t}{\partial \phi} = -\frac{\partial F}{\partial \phi}\bigg/\frac{\partial F}{\partial t} \tag{G.19}$$

and

$$\frac{\partial^2 t}{\partial \phi^2} = \left(-\frac{\partial F}{\partial t}\frac{\partial^2 F}{\partial \phi^2} + \frac{\partial F}{\partial \phi}\frac{\partial^2 F}{\partial \phi}\right)\bigg/\left(\frac{\partial F}{\partial t}\right)^2. \tag{G.20}$$

Evaluation of the derivatives at P_0 yields

$$\frac{\partial t}{\partial \phi}\bigg|_{P_0} = 0 \tag{G.21}$$

and

$$\left.\frac{\partial^2 t}{\partial \phi^2}\right|_{P_0} = \rho_0 \left.\frac{\partial \rho}{\partial t}\right|_{P_0}. \tag{G.22}$$

Finally we must evaluate the derivatives $\partial \phi / \partial l_1$ and $\partial \phi / \partial l_1^2$. The arc length referenced to the point P_0 is given by

$$l_1(\phi) = \int_0^\phi \left|\frac{d\mathbf{r}}{d\phi'}\right| d\phi'. \tag{G.23}$$

Using implicit differentiation on this formula we obtain

$$\frac{d\phi}{dl_1} = 1 \Big/ \left.\left|\frac{d\mathbf{r}}{d\phi'}\right|\right|_{\phi'=\phi} \tag{G.24}$$

where

$$\frac{d\mathbf{r}}{d\phi'} = \frac{\partial \mathbf{r}}{\partial \phi'} + \frac{dt}{d\phi'}\frac{\partial \mathbf{r}}{\partial t}. \tag{G.25}$$

Evaluating these expressions gives

$$\left.\frac{d\phi}{dl_1}\right|_{P_0} = \frac{1}{\rho_0}. \tag{G.26}$$

In a similar manner we may evaluate $d^2\phi/dl_1^2$, and determine

$$\left.\frac{d^2\phi}{dl_1^2}\right|_{P_0} = 0. \tag{G.27}$$

The required partial derivatives may now be written as

$$\left.\frac{\partial}{\partial l_1}\right|_{P_0} = \frac{\partial}{\rho_0 \partial \phi} \tag{G.28}$$

and

$$\left.\frac{\partial^2}{\partial l_1^2}\right|_{P_0} = \frac{1}{\rho_0^2}\frac{\partial^2}{\partial \phi^2} + \frac{1}{\rho_0}\left.\frac{d\rho}{dt}\right|_{P_0}\frac{\partial}{\partial t}. \tag{G.29}$$

The transformation of the mixed partial derivative is the last to be found. In order to determine this derivative we consider the situation depicted in Figure 7.5. We wish to determine the change in an arbitrary function after a small movement along the l_2 curve to the point P_1, and then a movement along the l_1 curve, as indicated on Figure 7.5. As before the derivative on the l_1 is given by

$$\frac{\partial}{\partial l_1} = \left(\frac{d\phi}{dl_1}\right)\left(\frac{\partial}{\partial \phi} + \frac{dt}{d\phi}\frac{\partial}{\partial t}\right). \tag{G.30}$$

In this case both $d\phi/dl_1$ and $dt/d\phi$ depend on the initial value of l_2. The mixed partial derivative is given by

$$\frac{\partial^2}{\partial l_1 \partial l_2} = \left(\frac{\partial^2 \phi}{\partial l_2 \partial l_1}\right)\left(\frac{\partial}{\partial \phi} + \frac{dt}{d\phi}\frac{\partial}{\partial t}\right)$$
$$+ \left(\frac{d\phi}{dl_1}\right)\left(\frac{\partial^2}{\partial \phi \partial l_2} + \left[\frac{\partial}{\partial l_2}\left(\frac{dt}{d\phi}\right)\right]\right) + \left(\frac{dt}{d\phi}\right)\frac{\partial^2}{\partial l_2 \partial t}\right) \qquad \text{(G.31)}$$

The necessary derivatives are obtained by generalizing Eqs. (G.17) and (G.18), including the fact that the l_1 curve now originates at a point determined by l_2,

$$F(t, \phi) = \left.\frac{\partial \rho}{\partial t}\right|_{P_0}(\rho(t)\cos\phi - \rho_1)) + \left.\frac{\partial z}{\partial t}\right|_{P_0}(z(y) - z_1) = 0. \qquad \text{(G.32)}$$

Using implicit differentiation we find that on the l_1 curve,

$$\frac{\partial t}{\partial \phi} = 0, \qquad \text{(G.33)}$$

for any value of l_2, and thus

$$\left[\frac{\partial}{\partial l_2}\frac{\partial t}{\partial \phi}\right] = 0. \qquad \text{(G.34)}$$

By generalizing the arc length results we find,

$$\frac{d\phi}{dl_1} = \frac{1}{\rho_1(l_2)} \qquad \text{(G.35)}$$

and thus

$$\frac{d^2\phi}{dl_2 dl_1} = \frac{-1}{(\rho_1(l_2))^2}\frac{d\rho_1}{dl_2} \qquad \text{(G.36)}$$

Evaluating all the expressions at the point $l_1 = l_2 = 0$ and plugging into Eq. (G.31) results in the following expression for the mixed partial derivative,

$$\left.\frac{\partial^2}{\partial l_1 \partial l_2}\right|_{P_0} = \frac{1}{\rho_0}\frac{\partial^2}{\partial \phi \partial t} - \frac{1}{\rho_0^2}\left.\frac{d\rho}{dt}\right|_{P_0}\frac{\partial}{\partial \phi} \qquad \text{(G.37)}$$

For completeness we rewrite the other four differential operators here,

$$\left.\frac{\partial}{\partial l_1}\right|_{P_0} = \frac{1}{\rho_0}\frac{\partial}{\partial \phi}, \qquad \text{(G.38)}$$

$$\left.\frac{\partial^2}{\partial l_1^2}\right|_{P_0} = \frac{1}{\rho_0^2}\frac{\partial^2}{\partial \phi^2} + \frac{1}{\rho_0}\left.\frac{d\rho}{dt}\right|_{P_0}\frac{\partial}{\partial t}, \qquad \text{(G.39)}$$

$$\frac{\partial}{\partial l_2}\bigg|_{P_0} = \frac{\partial}{\partial t},\tag{G.40}$$

and

$$\frac{\partial^2}{\partial l_2^2}\bigg|_{P_0} = \frac{\partial^2}{\partial t^2}.\tag{G.41}$$

Having transformed the operators and found the unit vectors, the differential Eqs. (7.24) and (7.25) may now be written in terms of ϕ and t and the vector components V_ϕ and V_t. For example, consider the operation $\partial^2 V_1/\partial l_1 \partial l_2$,

$$\frac{\partial^2 V_1}{\partial l_1 \partial l_2} = \left(\frac{1}{\rho}\frac{\partial^2}{\partial t \partial \phi} - \frac{1}{\rho^2}\frac{d\rho}{dt}\frac{\partial}{\partial \phi}\right)\left(V_t \hat{\mathbf{t}}\cdot \hat{\mathbf{a}}_1 + V_\phi \hat{\boldsymbol{\phi}}\cdot \hat{\mathbf{a}}_1\right).\tag{G.42}$$

Expanding the derivatives using the chain rule gives

$$\frac{\partial^2 V_1}{\partial l_1 \partial l_2}\tag{G.43}$$

$$= \frac{1}{\rho}\left[\frac{\partial^2 V_t}{\partial t \partial \phi}\left(\hat{\mathbf{t}}\cdot \hat{\mathbf{a}}_1\right) + \frac{\partial^2 \left(\hat{\mathbf{t}}\cdot \hat{\mathbf{a}}_1\right)}{\partial t \partial \phi}V_t + \frac{\partial V_t}{\partial \phi}\frac{\partial \left(\hat{\mathbf{t}}\cdot \hat{\mathbf{a}}_1\right)}{\partial t} + \frac{\partial V_t}{\partial t}\frac{\partial \left(\hat{\mathbf{t}}\cdot \hat{\mathbf{a}}_1\right)}{\partial \phi}\right.$$

$$\left. + \frac{\partial^2 V_\phi}{\partial t \partial \phi}\left(\hat{\boldsymbol{\phi}}\cdot \hat{\mathbf{a}}_1\right) + \frac{\partial^2 \left(\hat{\boldsymbol{\phi}}\cdot \hat{\mathbf{a}}_1\right)}{\partial t \partial \phi}V_\phi + \frac{\partial V_\phi}{\partial \phi}\frac{\partial \left(\hat{\boldsymbol{\phi}}\cdot \hat{\mathbf{a}}_1\right)}{\partial t} + \frac{\partial V_\phi}{\partial t}\frac{\partial \left(\hat{\boldsymbol{\phi}}\cdot \hat{\mathbf{a}}_1\right)}{\partial \phi}\right]$$

$$- \frac{1}{\rho^2}\frac{d\rho}{dt}\left[\frac{\partial V_t}{\partial \phi}\left(\hat{\mathbf{t}}\cdot \hat{\mathbf{a}}_1\right) + \frac{\partial \left(\hat{\mathbf{t}}\cdot \hat{\mathbf{a}}_1\right)}{\partial \phi}V_t + \frac{\partial V_\phi}{\partial \phi}\left(\hat{\boldsymbol{\phi}}\cdot \hat{\mathbf{a}}_1\right) + \frac{\partial \left(\hat{\boldsymbol{\phi}}\cdot \hat{\mathbf{a}}_1\right)}{\partial \phi}V_\phi\right]$$

Evaluating the derivatives gives the following final expression,

$$\frac{\partial^2 V_1}{\partial l_1 \partial l_2} = \frac{1}{\rho}\left[\frac{\partial^2 V_\phi}{\partial \phi \partial t} + \frac{d\rho}{dt}\frac{\partial V_t}{\partial t} - \frac{d\rho}{dt}\left(\frac{d\rho}{dt}\frac{d^2\rho}{dt^2} + \frac{dz}{dt}\frac{d^2 z}{dt^2}\right)V_t\right]$$

$$- \frac{1}{\rho^2}\frac{d\rho}{dt}\left(\frac{\partial V_\phi}{\partial \phi} + \frac{d\rho}{dt}V_t\right)\tag{G.44}$$

The rest of the expressions may be transformed in a similar manner, resulting in

$$\frac{\partial^2 V_2}{\partial l_1 \partial l_2} = \frac{1}{\rho}\left[\frac{\partial^2 V_t}{\partial \phi \partial t} - \frac{d^2\rho}{dt^2}V_\phi - \frac{d\rho}{dt}\frac{\partial V_\phi}{\partial t}\right] - \frac{1}{\rho^2}\frac{d\rho}{dt}\left(\frac{\partial V_t}{\partial \phi} - \frac{d\rho}{dt}V_\phi\right)\tag{G.45}$$

$$\frac{\partial^2 V_1}{\partial l_1^2} = \frac{1}{\rho^2}\left[\frac{\partial^2 V_\phi}{\partial \phi^2} - \left(\frac{d\rho}{dt}\right)^2 V_\phi + 2\frac{d\rho}{dt}\frac{\partial V_t}{\partial \phi} + \rho\frac{d\rho}{dt}\frac{\partial V_\phi}{\partial t}\right]\tag{G.46}$$

$$\frac{\partial^2 V_2}{\partial l_1^2} = \frac{1}{\rho^2} \left[\frac{\partial^2 V_t}{\partial \phi^2} - \left(\frac{d\rho}{dt}\right)^2 V_t - 2\frac{d\rho}{dt}\frac{\partial V_\phi}{\partial \phi} + \rho\frac{d\rho}{dt}\frac{\partial V_t}{\partial t} \right] \qquad \text{(G.47)}$$

$$\frac{\partial^2 V_1}{\partial l_2^2} = \frac{\partial^2 V_\phi}{\partial t^2} \qquad \text{(G.48)}$$

and

$$\frac{\partial^2 V_2}{\partial l_2^2} = \frac{\partial^2 V_t}{\partial t^2}. \qquad \text{(G.49)}$$

Appendix H

Matrix Elements for the HOIBC Portion of the BOR Solution

The matrix elements for $[P_1]$ through $[P_8]$ which appear in the system matrix for the HOIBC solution are somewhat complicated. Their entries are given by

$$P_1(i,j) = F_1(i,j) + c_2 n^2 F_2(i,j) - c_1 \left(F_3(i,j) + F_4(i,j) - F_5(i,j) \right) - F_8(i,j) \tag{H.1}$$

$$P_2(i,j) = jn \left(c_1 - c_2 \right) F_6(i,j) - jn \left(c_1 + c_2 \right) F_7(i,j) \tag{H.2}$$

$$P_3(i,j) = jn \left(c_1 - c_2 \right) F_6(i,j) + jn \left(c_1 + c_2 \right) F_7(i,j) \tag{H.3}$$

$$P_4(i,j) = F_1(i,j) + c_1 n^2 F_2(i,j) - c_2 \left(F_3(i,j) + F_4(i,j) - F_5(i,j) \right) + F_9(i,j) \tag{H.4}$$

$$P_5(i,j) = jn \left(c_4 - c_5 \right) F_6(i,j) - jn \left(c_4 + c_5 \right) F_7(i,j) \tag{H.5}$$

$$P_6(i,j) = -c_3 F_1(i,j) - c_5 n^2 F_2(i,j) + c_4 \left(F_3(i,j) + F_4(i,j) - F_5(i,j) \right) - F_9(i,j) \tag{H.6}$$

$$P_7(i,j) = c_3 F_1(i,j) + c_4 n^2 F_2(i,j) - c_5 \left(F_3(i,j) + F_4(i,j) - F_5(i,j) \right) + F_8(i,j) \tag{H.7}$$

167

and

$$P_8(i,j) = -jn\,(c_4 - c_5)\,F_6(i,j) - jn\,(c_4 + c_5)\,F_7(i,j) \tag{H.8}$$

The interaction integrals F_1 through F_9 are given by

$$F_1(i,j) = \int_{C_0} T_i(t)T_j(t)dt \tag{H.9}$$

$$F_2(i,j) = \int_{C_0} T_i(t)\frac{T_j(t)}{\rho(t)}dt \tag{H.10}$$

$$F_3(i,j) = \int_{C_0} T_i(t)\rho(t)\frac{d^2}{dt^2}\left(\frac{T_j(t)}{\rho(t)}\right)dt \tag{H.11}$$

$$F_4(i,j) = \int_{C_0} T_i(t)\frac{d\rho(t)}{dt}\frac{d}{dt}\left(\frac{T_j(t)}{\rho(t)}\right)dt \tag{H.12}$$

$$F_5(i,j) = \int_{C_0} T_i(t)\left(\frac{d\rho(t)}{dt}\right)^2\left(\frac{T_j(t)}{\rho(t)^2}\right)dt \tag{H.13}$$

$$F_6(i,j) = \int_{C_0} T_i(t)\frac{d}{dt}\left(\frac{T_j(t)}{\rho(t)}\right)dt \tag{H.14}$$

$$F_7(i,j) = \int_{C_0} T_i(t)\left(\frac{d\rho(t)}{dt}\right)\left(\frac{T_j(t)}{\rho(t)}\right)dt \tag{H.15}$$

$$F_8(i,j) = \int_{C_0} T_i(t)\left(\frac{d^2\rho(t)}{dt^2}\right)^2\left(\frac{T_j(t)}{\rho(t)}\right)dt \tag{H.16}$$

and

$$F_9(i,j) = \int_{C_0} T_i(t)\left(\frac{d\rho(t)}{dt}\frac{d^2\rho(t)}{dt^2} + \frac{dz(t)}{dt}\frac{d^2z(t)}{dt^2}\right)T_j(t)dt \tag{H.17}$$

All of the above integrals are found in closed form. The integrals F_8 and F_9 are poorly defined due to the facet approximation of the BOR. The second derivatives $d^2\rho/dt^2$ and d^2z/dt^2 are nonzero only at the junction points of the facets making up the BOR; however, $d\rho/dt$ and dz/dt are undefined at these points. To rectify this problem the first derivatives at a point joining two facets are taken to be equal to the average of their values on these two adjoining facets. Fortunately, it has been found that the contribution of these two interaction integrals to the overall MOM matrix is in general negligible.

Bibliography

[1] M. I. Skolnik, editor. *Radar Handbook.* McGraw-Hill, New York, 1970.

[2] P. S. Kildal, E. Lier, and J. A. Aas. Artificially hard and soft surfaces in electromagnetics and their application. In *Proceedings of the 1988 IEEE Antennas and Propagation Society Symposium*, pages 832–835, Syracuse, NY, June 1988.

[3] R. F. Harrington. *Field Computation by Moment Methods.* Macmillan, New York, 1968.

[4] R. C. Hansen, editor. *Geometric Theory of Diffraction.* IEEE Press, New York, 1981.

[5] P. Ufimstev. Method of edge waves in the physical theory of diffraction. *Izd-Sovyetskoye Radio*, pages 1–243, 1962.

[6] D. Duan, Y. Rahmat-Samii, and J. P. Mahon. Scattering from a circular disk: A comparative study of PTD and GTD techniques. *Proceedings of the IEEE*, 79(10):1472–1480, October 1991.

[7] P. L. Huddleston, L. N. Medgyesi-Mitschang, and J. M. Putnam. Combined field integral equation formulation for scattering by dielectrically coated conducting bodies. *IEEE Transactions on Antennas and Propagation*, 34(4):510–520, April 1986.

[8] J. M. Jin and V. Liepa. Application of hybrid finite element to electromagnetic scattering from coated cylinders. *IEEE Transactions on Antennas and Propagation*, 36(1):50–54, January 1988.

[9] R. Mittra, O. Ramahi, A. Khebir, R. Gordon, and A. Kouki. On the use of absorbing boundary conditions for electromagnetic scattering problems. *IEEE Transactions on Magnetics*, 25(7):3034–3040, July 1989.

[10] M. A. Leontovich. *Investigations on Radiowave Propagation, Part II.* Moscow: Academy of Sciences, 1948.

[11] S. W. Lee and W. Gee. How good is the impedance boundary condition? *IEEE Transactions on Antennas and Propagation*, 35(11):1313–1315, November 1987.

[12] N. G. Alexopoulos and G. A. Tadler. Accuracy of the Leontovich boundary condition for continuous and discontinuous surface impedance. *Journal of Applied Physics*, 46(8):3326–3332, August 1975.

[13] J. R. Rogers. Numerical solutions to ill-posed and well-posed impedance boundary condition integral equations. Technical Report 642, Massachusetts Institute of Technology, Lincoln Laboratory, Cambridge, Mass., November 1983.

[14] L. N. Medgyesi-Mitschang and J. M. Putnam. Integral equation formulations for imperfectly conducting scatters. *IEEE Transactions on Antennas and Propagation*, 33(2):206–214, February 1986.

[15] P. L. Huddleston. Scattering by finite open cylinders using approximate boundary conditions. *IEEE Transactions on Antennas and Propagation*, 37(2):253–257, February 1989.

[16] A. A. Kishk. Electromagnetic scattering from composite objects using a mixture of exact and impedance boundary conditions. *IEEE Transactions on Antennas and Propagation*, 39(6):826–833, June 1991.

[17] P. S. Kildal. Definition of artificially soft and hard surfaces for electromagnetic waves. *Electronics Letters*, 24:168–170, 1988.

[18] K. W. Whites. Scattering models for infinite periodic structures at low frequencies. In *Proceedings of the 1991 IEEE Antennas and Propagation Society Symposium*, pages 268–271, London, Ontario, Canada, June 1991.

[19] S. N. Karp and F. C. Karal Jr. Generalized impedance boundary conditions with applications to surface wave structures. In *Electromagnetic Theory Part 1*, pages 479–483. Pergamon, New York, 1965.

[20] A. L. Weinstein. *The Theory of Diffraction and the Factorization Method*. Golem, Boulder, Colorado, 1969.

[21] T. B. A. Senior and J. L. Volakis. Derivation and application of a class of generalized impedance boundary conditions. *IEEE Transactions on Antennas and Propagation*, 37(12):1566–1572, December 1989.

[22] M. A. Ricoy and J. L. Volakis. Derivation of generalized transition/boundary conditions for planar multiple-layer structures. *Radio Science*, 25(4):391–405, July 1990.

[23] M. Idemen. Universal boundary conditions of the electromagnetic field. *Journal of the Physical Society of Japan*, 59(1):71–80, January 1990.

[24] J. L. Volakis and T. B. A. Senior. Application of a class of generalized boundary conditions to scattering by a metal-backed dielectric half-plane. *Proceedings of the IEEE*, 77(5):796–805, May 1989.

[25] T. B. A. Senior and J. L. Volakis. Sheet simulation of a thin dielectric layer. *Radio Science*, 22(7):1261–1272, December 1987.

[26] K. Barkeshli and J. L. Volakis. TE scattering by a one-dimensional groove in a ground plane using higher order impedance boundary conditions. *IEEE Transactions on Antennas and Propagation*, 38(9):1421–1428, September 1990.

[27] J. L. Volakis and H. H. Syed. Application of higher order boundary conditions to scattering by multilayer coated cylinders. *Journal of Electromagnetic Waves and Applications*, 4(12):1157–1180, December 1990.

[28] D. L. Jaggard, A. R. Mickelson, and C. H. Papas. On electromagnetic waves in chiral media. *Applied Physics*, 18:211–216, 1979.

[29] R. D. Graglia, P. L. E. Uslengi, and R. E. Zich. Reflection and transmission for planar structures of bianisotropic media. *Electromagnetics*, 11:193–208, 1991.

[30] D. J. Hoppe and Y. Rahmat-Samii. Higher-order impedance boundary conditions for anisotropic and nonreciprocal coatings. In *Proceedings of the 1992 IEEE Antennas and Propagation Society Symposium*, pages 1993–1996, Chicago, Ill., July 1992.

[31] T. B. A. Senior. Generalized boundary conditions and the question of uniqueness. *Radio Science*, 27(6):929–934, November 1992.

[32] N. Engheta and D. L. Jaggard. Electromagnetic chirality and its applications. *IEEE Antennas and Propagation Society Newsletter*, 30:6–12, October 1988.

[33] V. K. Varadan, V. V. Varadan, and A. Lakhtakia. On the possibility of designing anti-reflection coatings using chiral composites. *Journal of Wave-Material Interaction*, 2(1):71–81, January 1987.

[34] D. L. Jaggard and J. R. Liu. Chiral layers for RCS control of curved surfaces. In *Proceedings of the 1991 URSI Radio Science Meeting*, page 351, London, Ontario, Canada, June 1991.

[35] S. Bassiri, N. Engheta, and C. H. Papas. Dyadic Green's function and dipole radiation in chiral media. *Alta Frequenza*, LV:83–88, March 1986.

[36] R. G. Rojas and L. M. Chou. Generalized impedance/resistive boundary conditions for a planar chiral slab. In *Proceedings of the 1991 URSI Radio Science Meeting*, page 358, London, Ontario, Canada, June 1991.

[37] R. F. Harrington. *Time-Harmonic Electromagnetic Fields*. McGraw-Hill, New York, 1961.

[38] M. S. Kluskens and E. H. Newman. Scattering by a multilayer chiral cylinder. *IEEE Transactions on Antennas and Propagation*, 39(1):91–96, January 1991.

[39] K. W. Whites, E. Michielssen, and R. Mittra. Approximating the scattering by a material-filled trough in an infinite plane using the impedance boundary condition. *IEEE Transactions on Antennas and Propagation*, 41(2):146–153, February 1993.

[40] R. A. Hurd. The edge condition in electromagnetics. *IEEE Transactions on Antennas and Propagation*, 24(1):70–73, January 1976.

[41] J. Meixner. The behavior of electromagnetic fields at edges. *IEEE Transactions on Antennas and Propagation*, 20(4):442–446, July 1972.

[42] G. L. James. Analysis and design of TE_{11}-to-HE_{11} corrugated cylindrical waveguide mode converters. *IEEE Transactions on Microwave Theory and Techniques*, 29(10):1059–1066, October 1981.

[43] D. J. Hoppe and Y. Rahmat-Samii. Scattering by superquadric dielectric-coated cylinders using higher-order impedance boundary conditions. *IEEE Transactions on Antennas and Propagation*, 40(12):1513–1523, December 1992.

[44] A. J. Poggio and E. K. Miller. Integral equation solutions of three-dimensional scattering problems. In *Computer Techniques for Electromagnetics*, chapter 4. Pergamon, New York, 1973.

[45] E. D. Constantinides and R. J. Marhefka. Plane wave scattering from 2-D perfectly conducting superquadric cylinders. *IEEE Transactions on Antennas and Propagation*, 39(3):367–376, March 1991.

[46] D. J. Hoppe and Y. Rahmat-Samii. Scattering by coated bodies of revolution using higher order impedance boundary conditions. In *Proceedings of the 1993 IEEE Antennas and Propagation Society Symposium*, pages 1764–1767, Ann Arbor, Mich., June 1993.

[47] N. Engheta and P. Pelet. Surface waves in chiral layers. *Optics Letters*, 16(10):723–725, May 1991.

Index

T - #0196 - 101024 - C0 - 229/152/10 [12] - CB - 9781560323853 - Gloss Lamination